D0216214

PRIMATE COMPARATIVE ANATOMY

Primate Comparative ANATOMY

DANIEL L. GEBO

Illustrations by Mat Severson

JOHNS HOPKINS UNIVERSITY PRESS | BALTIMORE

© 2014 Johns Hopkins University Press
All rights reserved. Published 2014
Printed in the United States of America on acid-free paper
9 8 7 6 5 4 3 2 1

Johns Hopkins University Press
2715 North Charles Street
Baltimore, Maryland 21218-4363
www.press.jhu.edu

Library of Congress Cataloging-in-Publication Data

Gebo, Daniel Lee, 1955–
    Primate comparative anatomy / Daniel L. Gebo ; with Illustrations
by Mat Severson.
        pages cm
    Includes bibliographical references and index.
    ISBN 978-1-4214-1489-8 (hardcover : alk. paper) — ISBN
978-1-4214-1490-4 (electronic) — ISBN 1-4214-1489-9 (hardcover :
alk. paper) — ISBN 1-4214-1490-2 (electronic) 1. Primates—Anatomy.
I. Title.
    QL737.P9G37 2014
    599.8147—dc23   2013046570

A catalog record for this book is available from the British Library.

*Special discounts are available for bulk purchases of this book. For
more information, please contact Special Sales at 410-516-6936 or
specialsales@press.jhu.edu.*

Johns Hopkins University Press uses environmentally friendly book
materials, including recycled text paper that is composed of at least
30 percent post-consumer waste, whenever possible.

To Percy, Katniss, and Peeta, fictional characters who captured my daughter's imagination and in the process taught her to enjoy reading. Live long and prosper, AMG.

CONTENTS

PREFACE

After teaching primate comparative anatomy for a number of years, I realized that this subject should be simpler for students and professionals alike. I wrote this volume to fit the great diversity of primate morphology among living primates into meaningful subunits. Any complaints about errors or overly simplified "morphological stories" can be directed toward me, as I have tried my best to streamline primate comparative anatomy in order to get a "take home" message across. There are plenty of specialized articles and books that can be examined after reading this one, given the ever growing banks of information on primate anatomy, evolution, ecology, and taxonomy.

Anatomy books are only as good as their illustrations, and I especially thank Mat Severson for all of his artistic efforts, hard work, and time over the past several years to complete this volume. It is our hope that these illustrations show primate anatomy realistically and that the images are accurate enough to be used next to actual bones. Although we have selected a subset of primates to illustrate, we hope that the selected taxa will offer a survey of primate anatomical diversity across the entire Order Primates.

This is a book about primate anatomy with a comparative focus and it surveys all families of living primates. It is largely focused on the skeletal system to comport with most classroom settings, but soft tissue anatomy, for example, muscles and brains, is discussed briefly when appropriate. I decided early on to avoid any lengthy discussions on soft tissue comparative anatomy as a way to simplify this volume. In a similar vein, I largely excluded comments on particular fossil primates to minimize both complexity and the details required to explain each fossil group sufficiently. As a compromise, I have written a small section on primate evolution in chapter 1 to anchor the living primates, and I added illustrated boxes on a variety of fossil primates to the individual chapters for a few notable comparisons with living primates, providing a little flavor for this interesting but broad theme.

This volume is largely written from my lecture notes over many years of teaching and I admit to having borrowed liberally from many sources. At the end of each chapter, I have listed important references that provided expertise or helped me explain particular anatomical points. I hope that this volume will be useful to undergraduate and early graduate students in setting the stage for advanced work in comparative anatomy. The comparative anatomy of primates is a resource for taxonomy, functional anatomy, and general primate biology, but this field also represents the backbone for paleoprimatology and in so doing is the foundation for understanding our past.

* * *

I wish to thank Johns Hopkins University Press and especially Vince Burke, my editor, for taking a chance on me and this volume. All of the JHUP staff were generous with their time. I especially thank Deborah Bors, Catherine Goldstead, Jen Malat, and my copy editor, Maria E. denBoer. I also thank the staff at the Field Museum of Natural History (Chicago) for their assistance with specimens throughout my many years at Northern Illinois University. I thank two anonymous reviewers for their many thoughtful and useful comments. Drs. Bert Covert, Marian Dagosto, Leila Porter, and Blythe Williams also provided a variety of helpful comments. I am grateful to Drs. John Fleagle and Rich Kay for permission to redraw and modify several figures. Last, I thank David Haring (Duke Lemur Center) and Stacy Frank (Minden Pictures) for providing beautiful images of living primates in chapters 2 and 3.

PRIMATE COMPARATIVE ANATOMY

1

Primate Phylogeny and Adaptation

The discovery of biological diversity early in human history allowed us to discriminate between organisms and to ultimately develop a classification of life. Although Aristotle appears to have been the first to classify the animal world in print (circa 384–322 BC), today we use a taxonomic system invented by Carolus Linnaeus from his famous volume *Systema Naturae*, published in 1758. Linnaeus's innovative system of binominal nomenclature uses only two words to name every organism: a genus and a species designation for all living and extinct species. All organisms are arranged hierarchically, from larger to smaller groupings, and they are organized by their comparative similarities. This initial effort to classify nature was based simply on comparative anatomy and it was performed long before our present-day understanding of evolution, genes, or DNA. All of these early attempts to classify animals, including primates, were ordered and categorical. It was Charles Darwin and his ideas concerning "descent with modification" that led to our current view that all living organisms are connected through genealogical descent. Modern classifications reflect our understanding of these evolutionary relationships. Thus, primate taxonomy reflects our current views concerning primate evolution. The taxonomy of living primates used in this volume is outlined in table 1.1.

PHYLOGENY

In contrast to taxonomy, the systematic ordering of taxa, phylogeny is the study of evolutionary lineages. Phylogeny places lineages, representing groups of organisms, within the tree of life. This is not always easy since many anatomical characters have evolved more than once and appear similar in design (evolutionary convergences). We want to identify homologous characters where anatomical shape stems from ancestry. There are many terms used in phylogeny and table 1.2 lists several important ones.

CLADOGRAMS

Cladograms arrange taxa by their evolutionary position. All cladograms arrange taxa at the top (fig. 1.1). This arrangement is not like an evolutionary tree and no fossil ancestors are recognized at the nodes. Cladograms are a curious way to look at phylogeny since we normally think of ancestor-descendant relationships as tree-like. In contrast to a phyletic tree, cladograms seek to identify taxa that are closely allied to each other. These sister taxa arrangements are aligned from the most primitive to the most advanced. Time is not reflected in a cladogram in any realistic sense (fig. 1.1). It is the changes in morphology—primitive to derived character states—that are shown. Although fossils can help determine character polarity (i.e., whether a feature is primitive or advanced), it is the changes in

Table 1.1 Primate Taxonomy

Kingdom	Animalia
Phylum	Chordata
Subphylum	Vertebrata
Class	Mammalia
Infraclass	Eutheria
Order	Primates
Suborder	Strepsirhini
Superfamily	Lemuroidea
Family	Cheirogaleidae
Family	Daubentoniidae
Family	Indriidae
Family	Lemuridae
Family	Lepilemuridae
Superfamily	Lorisoidea
Family	Lorisidae
Family	Galagidae
Suborder	Haplorhini
Infraorder	Tarsiiformes
Superfamily	Tarsioidea
Family	Tarsiidae
Infraorder	Simiiformes (or Anthropoidea)
Parvorder	Platyrrhini
Superfamily	Ceboidea
Family	Callitrichidae
Family	Cebidae
Family	Atelidae
Parvorder	Catarrhini
Superfamily	Cercopithecoidea
Family	Cercopithecidae
Superfamily	Hominoidea
Family	Hylobatidae
Family	Pongidae
Family	Hominidae

Table 1.2 Phylogenetic Terms

Phylogeny	Lineages, lines of descent, history
Homology	Structural similarity due to ancestry
Analogy	Functional similarities
Primitive characters	Characters shared by an ancestral group
Derived characters	Anatomical structures that define a group
Shared derived characters	Special anatomical structures that unite two taxa (synapomorphies)
Autapomorphic characters	Characters that are unique to a taxon or lineage
Cladistics	The study of clades or lineages of descent
Sister taxa	Two closely allied taxa
Cladogram	An array of sister taxa
Monophyletic	Referring to united taxa that share a common ancestry
Polarity of characters	The direction of character change through evolutionary lineages
Outgroup comparison	A comparison of closely related groups that helps to determine the polarity of traits
Convergence	Characters that are similar in structure or function but have separate origins
Parsimony	Adoption of the simplest explanation

morphology from one taxon to the next that reflect their true evolutionary position relative to each other. Any cladogram depicting an arrangement of primates represents a hypothesis to test the evolutionary relationships among taxa. Cladograms, over time, often become "true facts" concerning a primate's evolutionary position, such as the cladistic order of relationship of the living apes to humans.

Figure 1.2 shows three taxa (A, B, C). Which two are more closely related (i.e., are sister taxa)? There are several possibilities; in fact, there are four, if you consider the "I do not know" possibility (fig. 1.2d). This last possibility (fig. 1.2d) is called a trichotomy and it reflects the fact that we do not know which two of the three possible taxa are more closely related to each other. In the other cladistic arrangements (fig. 1.2a–c), we have identified a sister taxon relationship between the three taxa (A, B, C). For example,

Figure 1.2a shows A and C to be sister taxa relative to B. In each instance, one or more characters must be located at each of the nodes (N), or branching points. One taxon is viewed as more primitive than the other two, with sister taxa sharing at least one additional novel character. As the number of taxa and the list of morphological characters grow in size, we must use computer programs (i.e., PAUP or McClade) to help simplify the process of interpreting characters and sister taxa relationships through parsimony. Parsimony simply means the simplest explanation is best. In a cladistic analysis, parsimony helps us arrange taxa using the fewest number of morphological changes and reversals. In the end, cladograms are a good beginning in our thinking about evolution and the morphological transitions that must occur to sort through all of the living and extinct lineages of primates.

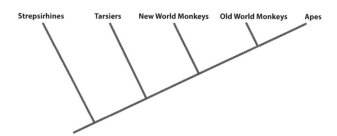

Figure 1.1 A primate cladogram with taxa arranged at the top.

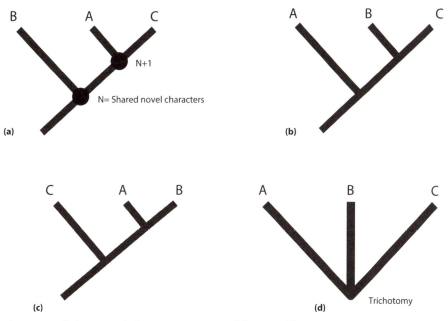

Figure 1.2 A cladogram with three taxa (A, B, C) and four possible outcomes.

GHOST LINEAGES

A common misunderstanding in evolutionary studies is that advanced forms cannot co-occur with their primitive ancestors. This is simply a misunderstanding of time and evolutionary relationships. For example, figure 1.3 shows two fossil "apes" that are found at approximately the same time (20 mya). One "ape" (*Morotopithecus*) is advanced, while the other (*Proconsul*) is primitive. Many wonder how this can happen in the fossil record if primitive forms are evolving into more advanced forms over time. In fact, this observation is made quite often in the fossil record as well as for living taxa. If the morphology of *Proconsul* is interpreted as being primitive, while the morphology of *Morotopithecus* is viewed as more advanced, this simply means that the common ancestry of *Proconsul* and *Morotopithecus* must extend farther back in time (before 20 mya). It does not require an evolutionary reversal from an advanced form back to a more primitive form. We call these more ancient connections, which are unknown in the fossil record at present, ghost lineages. If the polarity of morphological change between *Proconsul* and *Morotopithecus* is correct, then these two taxa must share a ghost lineage that goes back into the Oligocene. We could then go in search for a primitive *Proconsul*-like form earlier in time that would connect these two taxa.

ADAPTATION

A primate's survival is reflected in its behavioral ecology, how it uses its environment, its anatomy, and its evolutionary history. A simple triangular model (fig. 1.4) illustrates this relationship. Behavior, anatomy, and the environment are at the corners of the triangle. Behavior affects anatomy and the environment, while the environment affects anatomy and behavior. Evolutionary history affects all three. For example, a behavior such as locomotion determines where a primate moves and how it is capable of moving its body. Most of these abilities are the result of a long history of adaptation to enhance survival in a given, usually arboreal, environment. Heads, teeth, and bodies are thus mosaic structures that reflect a species' evolutionary past as well as their current survival abilities. Consider, for example, the body of humans as a series of upper body adaptations that reflect our arm-swinging ancestral past, while our hips and legs reflect a more recent evolutionary modification for bipedalism (locomotion by means of two hindlimbs; see chapter 10). The same is true for our teeth and our heads (fig. 1.5). In short, primate and human bodies are true time capsules of our ancestral past.

Understanding primate adaptations is important since this information allows us to comprehend survival abilities and it helps us to sequence the morphological changes that explain primate phylogeny. Adaptation and phylogeny go hand in hand in the science of primatology. By definition, an adaptation is a characteristic that allows an organism to survive and to reproduce in its environment. A niche is an organism's way to make a living; in other words, a niche is how an organism finds the resources needed to survive and compete against other organisms. When it comes to species, we often seek to examine how a single group is creatively modified into an array of different forms. We call this species explosion an adaptive radiation, and it means that closely related organisms have evolved to exploit different ecological niches. Adaptive radiations are the heart and soul of biology.

The competitive exclusion principle in ecology states that no two species may occupy the same ecological niche. If they do, competition is increased and one taxon will go extinct. Thus, organisms try to minimize competition. One of the best examples of this is documented for primates. Pierre Charles-Dominique (Muséum National d'Histoire Naturelle, Brunoy, France) collected dietary information on five nocturnal strepsirhines living in a West African forest (Gabon). Three species were galagos and two were lorises. All five of these primates are closely related, especially within each subgroup of lorises or galagos.

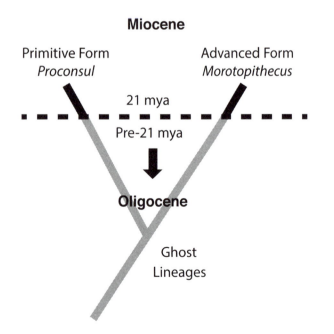

Figure 1.3 Advanced and primitive fossil "apes" living at the same time with ghost lineages connecting them farther back in time.

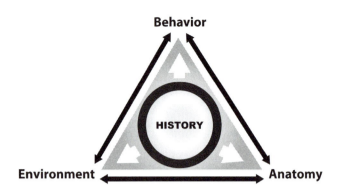

Figure 1.4 A simple triangular model of primate adaptation.

Since primates have many similar adaptations (i.e., similar teeth, guts, patterns of movements, etc.), how can five similarly adapted primates minimize competition and still co-exist? The answer lies in how they exploit different food items (table 1.3) and how they use the forest canopy.

First, all five species are nocturnal and thus they avoid monkeys that are active during the day in this West African forest. Each of the three galago species eats gums, fruit, or insects, but each has opted to specialize in consuming a specific food type (table 1.3): *Euoticus elegantulus* prefers gums, *Galago demidovii* eats insects, and *Galago alleni* consumes fruit. The two lorises, *Perodicticus potto* and *Arctocebus calabarensis*, have also chosen specialized diets. *Perodicticus* prefers fruit, while *Arctocebus* eats insects. Both *Galago demidovii* and *Arctocebus calabarensis* prefer to consume insects, but galagos eat active insects like grasshoppers and they are adept leapers, allowing them to hunt down and grab active insects. In contrast, *Arctocebus* eats well-hidden or camouflaged insects that often contain toxins. *Arctocebus* is a slow-moving, non-leaping primate that prefers to be cryptic in its use of the low understory and tangled undergrowth of the forest. In contrast, *Galago demidovii* prefers the upper canopy. Thus, the two insect-eating primates utilize different levels of the forest and prefer different types of insects, thereby minimizing competition. In a similar manner, the slow and cautious *Perodicticus* prefers to move within the high canopy to find its fruit, while *Galago alleni* prefers to leap between vertical supports among the trunks of trees below the canopy. Food consumption at different forest levels separates these two frugivorous primates, once again minimiz-

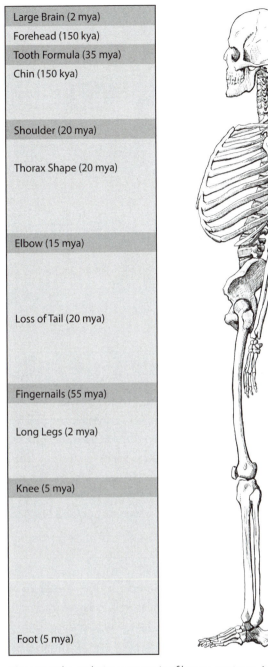

Large Brain (2 mya)

Forehead (150 kya)

Tooth Formula (35 mya)

Chin (150 kya)

Shoulder (20 mya)

Thorax Shape (20 mya)

Elbow (15 mya)

Loss of Tail (20 mya)

Fingernails (55 mya)

Long Legs (2 mya)

Knee (5 mya)

Foot (5 mya)

Figure 1.5 The evolutionary mosaic of human anatomy. Adapted from Fleagle, 1999.

ing food competition. *Euoticus elegantulus* is a specialized galago that often feeds on large-diameter vertical trunks. To accomplish this task, *Euoticus* evolved claw-like nails to cling to large-diameter tree trunks as it looks for gums to consume. Small primates with grasping hands and feet cannot hold on to large-diameter supports. As large-diameter, curved surfaces increase in size, hand and foot holds become less

Table 1.3 West African Galago and Loris Diets

	Gums (%)	Fruits (%)	Insects (%)
Galagos			
Galago demidovii	10	19	70
Euoticus elegantulus	75	3	20
Galago (Sciurocheirus) alleni	0	73	25
Lorises			
Perodicticus potto	21	65	10
Arctocebus calabarensis	0	14	85

Data from Charles-Dominique, 1977.

efficient on increasingly larger curved surfaces relative to their body size, grasp span, and grasping capabilities. Only because of its claws can *Euoticus* effectively exploit this food source, relative to the other West African primates, making it a specialized gumnivore.

BODY SIZE

Living primates range from 30 g (a pygmy mouse lemur, *Microcebus berthae*) to more than 200 kg (male gorillas, *Gorilla gorilla*). Some fossil primates have been estimated to have weighed as little as 10 g, while the fossil ape *Gigantopithecus* is thought to have been twice the size of present-day male gorillas. This makes the Order Primates diverse in terms of overall body sizes. The distribution of living species is skewed toward smaller-sized taxa weighing less than 5 kg, however. In general, strepsirhines are smaller than simiiforms, and terrestrial species are larger than arboreal ones. The largest arboreal mammal is the orangutan at 79 kg for an average male. Many primates are sexually dimorphic, with one sex being larger than the other. Old World monkeys and the great apes demonstrate the largest size differentials between the sexes.

Body size plays an especially important role in primate lives and in their adaptations. Size is a fundamental constraint in life history traits, diet, movement patterns, and other important ecological pressures such as predation. Body size influences an animal's metabolism. Kleiber's Law shows that the basal metabolic rate (or the energy consumed during the day while at rest) scales to body mass (weight$^{0.75}$). Metabolic rate is vastly higher for small taxa. For example, 10 g shrews starve within a 9-hour time period if not fed and 3 g shrews consume 125% of their body

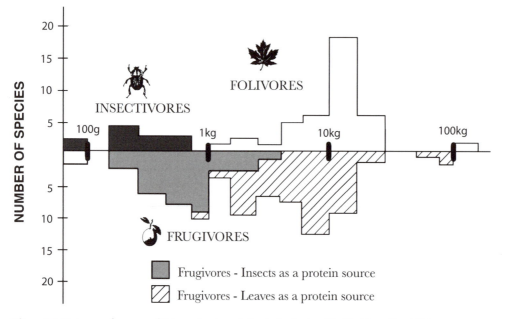

BODY WEIGHT AND DIET

Figure 1.6 Dietary preferences of living primates relative to body size. Modified from Kay, 1984.

weight daily! Although larger animals need a greater amount of food during any given day, large mammals consume far less nutritious (or more low-quality) food items to maintain their lower metabolic rate than do smaller mammals. In other words, large mammals can be more selective in their food choices, while small mammals need to find high-quality food items quickly and often. Richard Kay of Duke University has plotted primate dietary specialization relative to body mass. He shows that insectivorous primates are the smallest primates as they balance their high metabolic rate with the number of food items (i.e., insects) that can be caught and consumed each day, while leaf-eating primates are the largest. Fruit-eating primates overlap both the insectivorous and folivorous dietary groups. Insects and leaves provide the protein and fruits supply simple carbohydrates (fig. 1.6). Larger primates tend to have greater energy stores than their smaller cousins and this implies that they are better able to survive periods of food scarcity or in more variable resource environments.

Besides metabolism and diet, body size affects life history traits, with larger primates having slower and longer growth and developmental time periods. Predation tends to be lower with increased size. This

observation goes hand in hand with a slower birth rate for larger primates since all species need a balance between their reproductive output and rate of mortality. Table 1.4 compares the life history traits for a tiny primate, a mouse lemur, relative to that of a large ape, an orangutan, and this comparison shows a stark contrast in living styles, reproductive output, and lifespan.

WHAT IS A PRIMATE?

Primates, as a group, are characterized by a suite of anatomical and behavioral features. Relative to other mammals, primates' hands, feet, and eyes are more important characters. Primates possess grasping hands and feet with five digits and opposable thumbs and big toes. They possess nails instead of claws or hoofs. Primates have long limbs with good joint mobility, and they utilize an odd gait called diagonal couplets that keeps all but one limb attached to a branch when moving (see chapter 8). This gait sequence helps with balance along curved arboreal supports, especially small branches.

The primate brain is enlarged relative to those of most other mammals at any given size, making primates smart, creative, and flexible problem-solving

Table 1.4 Life History Traits

	Microcebus murinus	*Pongo pygmaeus*
Body size	62.3 g	36 kg, F; 83 kg, M
Mortality	25% per year	Low, < 1%
Diet	Insects, fruit, gums	Fruit, leaves, bark
Gestation length	60 days	250 days
Litter size	2	1
Interbirth interval	6 months	7 years
Offspring per lifetime	56	3
Age at weaning	40 days	8 years
Age at sexual maturity	8 months	15 years
Lifespan	15 years	60 years

mammals. One fundamental primate adaptation is brain related and involves their eyes. The forward shift of the orbits allows visual field overlap, which improves optic quality by aligning the frontal visual axis with the axis of each eye's lens. The overlapping visual fields allow primates to possess stereoscopic vision, or three-dimensional vision, a significant ability for animals moving and leaping high in the canopy. Besides forward-shifted orbits and eyes, primate brains have been rewired for vision enhancement. Not only do primates have crossing optical nerves (contralateral fibers), but they also have added optic connections on the same side of each eye (ipsilateral fibers; see chapter 5). All of this neuronal rewiring and brain expansion indicates that the occipital lobes of the primate brain must process all four visual fields from the right and left eyes in both the left and right occipital lobes to allow normal vision to occur. In short, primate vision is a complicated, computer-like synthesis that allows three-dimensional graphics to represent the outside world relative to the normal two-dimensional world of other mammals.

Primates also possess unique dental features and aspects of sociality that tend to distinguish them from other mammals. Primates have distinctive teeth with specific cusp patterns with a primitive dental formula of 2–1–4–3 (see chapter 6). Although some primate species are generally solitary in nature, most are not and are often quite social, sometimes with complex behavioral patterns that might include dominance, territoriality, and leadership roles. Primates often spend a great deal of time playing and interacting with one another. Primates have extended infant care, strong mother-infant bonds, and increased longevity. All of these behavioral traits suggest that learning is fundamental to primate survival. Primates, being relatively smart mammals, make learning a priority in their efforts to understand the environment they inhabit and this knowledge allows primates to compete for resources relative to other sympatric animals. Primates need to be competitive since they have lowered offspring per mating episodes relative to other mammals, making them K-selected. This biological designation means that primates invest heavily in a few offspring and survive due to their competitive abilities. In contrast, r-selected species survive largely on pure reproductive output.

Although Wilfred E. Le Gros Clark of Oxford University (United Kingdom) explained that tree living, or arboreality, shaped primate bodies and their special characteristics, this idea suffers from the fact that many mammals are arboreal and yet none look especially like primates. Thus, tree life by itself cannot explain why primates evolved the characteristics they all share. From this observation, Matt Cartmill (Boston University) developed the visual predation theory, which states that primates developed adaptations for tree life to prey on and catch quick and active insects. Primates are visually directed predators like owls and cats. Grasping, mobile limbs, diagonal couplets, and stereoscopic vision are adaptations that evolved to allow primates to move cautiously in the outer tree edges along slender branches to look for and catch insects as well as to consume flowers, nuts, or fruits located at the tips of these branches. Primates catch insect prey with their hands and bring these items back to their mouths for consumption. Thus, it is the activity of visual predation in trees that shaped the ancestral bodies of primates to look different from those of squirrels, palm civets, and other arboreal mammals.

In the fossil record, we find primates with all of the morphological changes noted above in the Early Eocene (55 mya) in two distinct primate lineages, the Adapiformes and the Tarsiiformes. Obviously, primate origins must occur before the Early Eocene. No clear transitional fossils have yet been found to document the forward placement of the orbits, brain expansion, or the evolution of nails. The fossil plesiadapiforms are often viewed as "archaic primates." They share several morphological features with primates, but they lack the key features noted above. Thus, there is a large morphological gap between Paleocene plesiadapiforms and primates of the Early Eocene. The closest living mammals to primates are the tree shrews (Order Scandentia) and the flying lemurs (Order Dermoptera), both from Asia. Neither morphological nor genetic analyses have been able to identify a consistent sister group for primates, and therefore it is unclear which of these three groups, the plesiadapiforms, tree shrews, or flying lemurs, is the true sister taxon to primates. Although there is not a majority view on primate origins at present, work is progressing rapidly in this area.

PRIMATE EVOLUTIONARY HISTORY

Although this volume focuses on the comparative anatomy and function of living primates, it is helpful to include a brief overview of primate evolutionary history. The Paleocene in the Early Tertiary is devoid of true primates with perhaps one exception (*Altiatlasius*) in Morocco. At the beginning of the Early Eocene, 55 million years ago, the primate story is completely changed with both modern suborders, Strepsirhini and Haplorhini, present and showing the hallmark adaptations of the living primates discussed above. For example, the orbits have moved forward, postorbital bars (see chapter 5) are present, legs are long for leaping, and hands and feet have been adapted for grasping, to name but a few classic primate characteristics. Fossil primates can be identified as Strepsirhini (adapiforms) or Haplorhini (tarsiiforms) by their leg anatomy. These two primate radiations represent the beginnings of primate evolution, especially their most basal members, and as such are the earliest lineages of primitive primates, often called prosimians in the past. The other major lineage of primates, Anthropoidea, a branch of Haplorhini, has no currently known fossil representatives in the Early Eocene. The most basal anthropoids are Middle Eocene in age (45 mya). All extant South American monkeys, Old World monkeys, and apes subsequently evolved from these basal anthropoids. In terms of geological time, anthropoid radiations span the Oligocene time epoch (35–22.5 mya), while the golden radiation of ape evolution is in the Miocene (22.5–5 mya).

Connecting fossil primates to specific living lineages of primates is often challenging in a specific sense, but we have been more successful in a broader view. The adapiform fossil primates, spanning six subfamilies, the Notharctinae, Cercamoniinae, Asiadapinae, Adapinae, Hoanghoninae, and Sivaladapinae, are the sister group of the toothcomb primates, Lemuriformes (fig. 1.7). Adapiforms occurred originally throughout the northern continents of Laurasia, North America, Europe, and Asia, and eventually they migrated to Africa, a southern Gondwanan continent. Adapiforms have never been found in South America, a continent strictly populated by platyrrhine monkeys that date from the latest Oligocene (i.e., *Branisella*). The adapiform taxa from North America and Europe have been known for more than a century and are the subjects of classic works in the field of primate evolution. The Asian and African adapiforms are becoming better known. *Adapis parisiensis* from France was the first named fossil primate, described by George Cuvier in 1821. *Notharctus* is the icon of North American adapiforms, having been featured in the classic monograph by William King Gregory in 1920. Fossil adapiforms are generally lemur-like in size, morphology, and functional abilities, being similar to *Notharctus*. They were diurnal fruit and leaf eaters with arboreal leaping tendencies. *Adapis* is a different type of an adapiform with a body anatomy similar in many ways to that of lorises, non-leaping arboreal primates. This mimicking of loris morphology is even more extreme in the Asian *Adapoides* or the African *Afradapis*.

Adapiforms are the most diverse in Europe and taxa such as *Donrussellia* (Europe), *Cantius* (Europe and North America), *Marcgodinotius* (Asia), and *Protoadapis* (Europe) represent primitive adapiforms. There is one character that clearly distinguishes the early members of this fossil group from living strepsirhines: four premolars. Living lemurs, lorises, and galagos possess three, one less than their ancestors. The best fossil evidence for toothcomb primates (Lemuriformes) is from the Late Eocene of the Fayum, Egypt, where an incisor of a toothcomb has been recovered for the taxon named *Saharagalago*, a primitive galagid. No other fossil primates are particularly closely related to the living families of Malagasy lemurs, the Lemuridae, Lepilemuridae, Daubentoniidae, Indriidae, and Cheirogaleidae. The African Galagidae and the

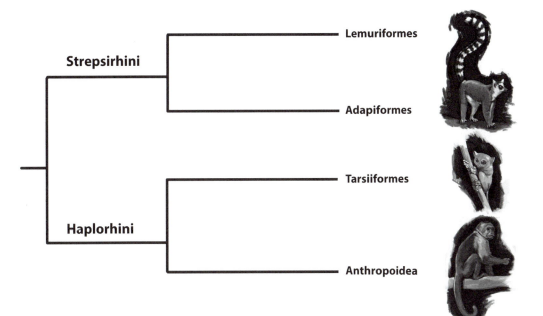

Figure 1.7 The evolutionary relationships of primate lineages.

African and Asian Lorisidae have a better fossil record with species in the Miocene of East Africa and India.

The fossil record of haplorhine (fig. 1.7) primates includes primitive tarsiiforms, the fossil families Archicebidae, Omomyidae, Microchoeridae, and Tarsiidae, as well the anthropoid radiations beginning with the Eosimiidae, a basal anthropoid lineage. The Archicebidae and the Omomyidae represent the most primitive of the tarsiiforms and the only fossils dated to the Early Eocene. Taxa such as *Archicebus, Teilhardina, Steinius,* and *Bataaromomys* represent primitive haplorhines. Both the archicebids and omomyids were generally small-sized primates, often mouse lemur–sized, with good leaping and climbing capabilities. They preferred to feed on insects and fruit. Several taxa within the Omomyidae possessed enlarged orbits for a nocturnal activity cycle, although nocturnality does not appear to be the case for the most primitive haplorhines, *Archicebus* or *Teilhardina*. *Archicebus* is known only from Asia, while omomyids are diverse in North America and are also found in Europe and Asia. The microchoerid radiation is best known in Europe with perhaps a few possible Asian fossils. European microchoerids are Middle Eocene in age and less diverse relative to contemporary European adapiforms. The tarsiids are known from Middle Eocene fossils from Asia and Miocene-age fossils from Southeast

Asia; this lineage has a living genus, *Tarsius*, with multiple species inhabiting many islands in Southeast Asia today. This is our best fossil link with a living taxon.

Anthropoid, or higher primate, evolution appears to have begun in Asia in the Middle Eocene with the Eosimiidae and subsequent lineages appearing in Asia and North Africa in the Late Eocene. The Eosimiidae are truly a stem taxon in that they possess a primitive mixture of omomyid-like features combined with a few anthropoid characteristics such as a tall mandible, obliquely oriented premolars, and foot joints resembling those of anthropoids. Eosimiids are tiny fossil primates ranging in size from 10–15 g to 150 g, a size range equivalent to that of many omomyids. At this body size, eosimiids likely consumed insects and fruit and are interpreted to be frequent leaping primates.

The Amphipithecidae have a long history as an enigmatic anthropoid group from Asia. Known from Myanmar and Thailand, *Amphipithecus* and *Pondaungia*, described as early as 1927 and 1937, and later taxa, such as *Ganlea, Myanmarpithecus*, and *Siamopithecus*, have been interpreted as early anthropoids, perhaps early catarrhines, or even as adapiform fossil primates. All are Late to Middle Eocene in age. The prevalent view is that these species are anthropoids, although which type is still debated. All amphipithecids are monkey-sized with odd dental features that have

Paleoprimatologists often use their anatomical knowledge of living primates to create reconstructions of extinct fossil primates. This image was created for a 55-million-year-old fossil primate named *Archicebus achilles*. An image like this requires a close dialogue between artist and scientist. Anatomical details such as body size, limb proportions, and tail length must be accurately drawn, while with items such as hair color there can be some "artistic license" taken. The same occurs for the environment and lifestyle of a fossil. In the end, good reconstructions should highlight and explain, as best as we may infer, how a fossil might actually have looked and behaved in the distant past.

Archicebus

the Proteopithecidae, are advanced relative to basal anthropoids, like the eosimiids, possessing many more anthropoid characters, including postorbital closure, hypocone development, and platyrrhine, monkey-like bodies. These two families represent monkey-sized taxa adapted for eating fruit and leaves and for arboreal quadrupedalism (locomotion by means of four limbs). Perhaps the most unusual limb anatomy is found in *Apidium*, a parapithecid, which has a long hindlimb with a lengthy tibia and a closely apposed fibula, adaptations for leaping, but with a foot containing short foot bones (or short "out levers"; see chapter 4), an odd combination for a leaping primate.

The oligopithecids and propliopithecids possess only two premolars like other catarrhines and generally resemble large platyrrhines in terms of their body anatomy. They represent stem catarrhines. Propliopithecids and oligopithecids are not great leapers but are well adapted for arboreal quadrupedalism, a typical anthropoid locomotor pattern. *Aegyptopithecus* and *Propliopithecus*, propliopithecids, have five cusped lower molars, the Y-5 pattern (see chapter 6), commonly observed among apes. In propliopithecid ear anatomy, the tympanic bone forms a ring structure similar to that of platyrrhines (see chapter 5) but not a tubular ectotympanic common among living Old World monkeys and apes (living catarrhine primates). *Saadanius*, an Oligocene fossil from Saudi Arabia, has a facial anatomy similar to that of *Aegyptopithecus*, but this taxon also has a tubular ectotympanic ear region, thereby demonstrating a closer evolutionary connection to living catarrhines.

The primate fossil record for South America is sparse initially with *Branisella* representing the oldest dental evidence, dating to the Late Oligocene (25–26 mya). The Miocene of South America has an array of fossil taxa, however, many fitting within living families of platyrrhines. Taxa such as *Caipota, Chilecebus, Dolichocebus, Homunculus, Lagonimico, Paralouatta, Protopithecus,* and *Tremacebus* are known from relatively complete skulls. In terms of adaptations, they span a wide array of dietary preferences and movements, as do living platyrrhines.

The Old World monkey fossil record begins in the Late Oligocene and Early to Middle Miocene of Africa with four taxa, *Nsungwepithecus, Prohylobates, Zaltanpithecus,* and *Victoriapithecus*. The Cercopithecidae are

been linked to feeding on hard objects, an ecological similarity to living pitheciine monkeys from South America.

The Fayum localities in Egypt have produced an abundance of anthropoid fossils and lineages, including the Parapithecidae, Proteopithecidae, Oligopithecidae, and Propliopithecidae. All of these families document early anthropoid or early catarrhine evolution. These four lineages are monkey-like in their overall anatomy and size. Even the two primitive families at the Fayum, the Parapithecidae and distinguished by their bilophodont molar teeth. *Victoriapithecus* and *Prohylobates* have this dental feature, albeit a more primitive version with an incomplete loph on the upper molars and small hypoconulids on the lower molars. The skull and postcranial elements of *Victoriapithecus* demonstrate a head and body anatomy similar to that of living cercopithecids. In fact, living and fossil cercopithecids are surprisingly similar in their anatomical form relative to other primate families. The cercopithecid lineage has reduced joint mobility across many limbs, suggesting a terrestrial adaptation and origin for this family. Many other cercopithecine and colobine fossil primates are known later in the Miocene, Pliocene, and Pleistocene paleontological records of Africa, Europe, and Asia.

Ape, or hominoid, evolution began in Africa generally in the Early Miocene with many primitive taxa, the best known being *Proconsul*, but recently a fossil "ape" from Tanzania was described from the Late Oligocene (25.2 mya), moving the start of the ape lineage back several million years. These fossil or "dental apes" with monkey-like bodies proliferate into 30 or more genera and some 50 African species throughout the Miocene. The Miocene epoch is the "Great Radiation of Apes." Eventually, apes migrated to Europe (e.g., *Dryopithecus* and *Oreopithecus*) and to Asia (e.g., *Sivapithecus* and *Gigantopithecus*). The primitive fossil apes with monkey-like bodies went extinct and Old World monkeys dominate modern tropical forests of Africa and Asia today. The only ape lineage to survive the changing environments of the Miocene and competition with Old World monkeys is the lineage that evolved an unusual upper body anatomy (see chapter 7) adapted for arm suspension and brachiation (arboreal locomotion by means of arm swinging). This lineage includes humans. Living African apes, having retreated from trees to become terrestrial quadrupeds or knuckle walkers, still possess a brachiating-adapted upper body, as do bipedal humans. This novel upper body anatomy is evident in certain European fossils such as *Dryopithecus, Oreopithecus*, and *Pierolapithecus* as well. Fossil apes, especially the great apes, evolved large body sizes relative to those of most monkeys. Many are more folivorous, although most apes still prefer fruit, their ancestral dietary preference. Size increases and decreases (e.g., gibbons) have also influenced ape locomotor adaptations across living and extinct forms. The human fossil record is rooted next to that of the African apes (see chapter 10).

Selected References

Allman, J. 1977. Evolution of the visual system in the early primates; pp. 1–53 *in* J.M. Epstein and A.N. Epstein (eds.), Progress in Psychobiology, Physiology, and Psychology, Vol. 7. Academic Press, New York.

Ankel-Simons, F. 2000. Primate Anatomy—An Introduction. 2nd edition. Academic Press, New York.

Beard, K.C., L. Marivaux, Y. Chaimanee, J.J. Jaeger, B. Marandat, P. Tafforeau, A.N. Soe, S.T. Tun, and A.A. Kyaw. 2009. A new primate from the Eocene Pondaung Formation of Myanmar and the monophyly of Burmese amphipithecids. Proceedings of the Royal Society of London (B) 276:3285–3294.

Carroll, R.L. 1988. Vertebrate Paleontology and Evolution. W.H. Freeman and Company, New York.

Cartmill, M. 1972. Arboreal adaptations and the origin of the Order Primates; pp. 97–122 *in* R.H. Tuttle (ed.), The Functional and Evolutionary Biology of Primates. Aldine Press, Chicago.

———. 1974. Pads and claws in arboreal locomotion; pp. 45–83 *in* J.A. Jenkins (ed.), Primate Locomotion. Academic Press, New York.

———. 1979. The volar skin of primates: its frictional characteristics and their functional significance. American Journal of Physical Anthropology 50:497–510.

Charles-Dominique, P. 1977. Ecology and Behaviour of Nocturnal Primates. Columbia University Press, New York.

Ciochon, R.L., and A.B. Chiarelli. 1980. Evolutionary Biology of the New World Monkeys and Continental Drift. Plenum Press, New York.

Conroy, G.C. 1990. Primate Evolution. W.W. Norton and Company, New York.

Cracraft, J., and N. Eldredge (eds.). 1979. Phylogenetic Analysis and Paleontology. Columbia University Press, New York.

Cuvier, G. 1821. Discours sur la Theorie de la Terre, Servant d'Introduction aux recherches sur les Ossements Fossiles. Paris.

Dagosto, M. 1993. Postcranial anatomy and locomotor behavior in Eocene primates; pp. 199–219 *in* D.L. Gebo (ed.), Postcranial Adaptation in Nonhuman Primates. Northern Illinois University Press, DeKalb.

Darwin, C. 1859. The Origin of Species by Means of Natural Selection or the Preservation of Favoured Races in the Struggle for Life. 1962 edition. Collier Macmillan Publishers, London.

Fleagle, J.G. 1999. Primate Adaptation and Evolution. Academic Press, New York.

Gause, G.F. 1934. The Struggle for Existence. Zoological Institute, University of Moscow, Moscow.

Gebo, D.L. (ed.). 1993. Postcranial Adaptation in Nonhuman Primates. Northern Illinois University Press, DeKalb.

———. 2004. A shrew-sized origin for primates. Yearbook of Physical Anthropology 47:40–62.

Gebo, D.L., L. MacLatchy, R. Kitio, A. Deino, J. Kingston, and D. Pilbeam. 1997. A hominoid genus from the Early Miocene of Uganda. Science 276:401–404.

Gregory, W.K. 1920. On the structure and relations of Notharctus: an American Eocene primate. Memoirs of the American Museum of Natural History 3:327–346.

Groves, C. 2001. Primate Taxonomy. Smithsonian Institution Press, Washington, D.C.

Hartwig, W.C. 2002. The Primate Fossil Record. Cambridge University Press, Cambridge, UK.

Hennig, W. 1966. Phylogenetic Systematics. University of Illinois Press, Champaign.

Jacobs, G.H. 1977. Color vision polymorphism in New World monkeys: implications for the evolution of primate trichromacy; pp. 45–74 in W.G. Kinzey (ed.), New World Primates: Ecology, Evolution and Behavior. De Gruyter, New York.

Jaeger, J.J., T. Thein, M. Benammi, Y. Chaimanee, A.N. Soe, T. Lwin, T. Tun, S. Wai, and S. Ducroq. 1999. A new primate from the Middle Eocene of Myanmar and the Asia early origin of anthropoids. Science 286:528–530.

Kay, R.F. 1984. On the use of anatomical features to infer foraging behavior in extinct primates; pp. 21–53 in J. Cant and P. Rodman (eds.), Adaptations for Foraging in Nonhuman Primates. Columbia University Press, New York.

Kimbel, W.H., and L.B. Martin (eds.). 1993. Species, Species Concepts, and Primate Evolution. Plenum Press, New York.

Kitching, I.J., P.L. Forey, C.J. Humphries, and D.M. Williams (eds.). 1998. Cladistics—The Theory and Practice of Parsimony Analysis. 2nd edition. Oxford University Press, Oxford, UK.

Kleiber, M. 1961. The Fire of Life—An Introduction to Animal Energetics. John Wiley and Sons, New York.

Larson, S.G. 1998. Unique aspects of quadrupedal locomotion in nonhuman primates; pp. 157–173 in E. Strasser, J.G. Fleagle, A.L. Rosenberger, and H.M. McHenry (eds.), Primate Locomotion: Recent Advances. Plenum Press, New York.

Le Gros Clark, W.E. 1959. The Antecedents of Man: An Introduction to the Evolution of Primates. University of Edinburgh Press, Edinburgh.

Linnaeus, C. 1758. Systema Naturae per Regna Tria Naturae, Secundam Classes, Ordines, Genera, Species cum Characteribus, Synonymis, Locis. Laurentii Sylvii, Stockholm.

Martin, R.D. 1990. Primate Origins and Evolution—A Phylogenetic Reconstruction. Princeton University Press, Princeton, N.J.

Morbeck, M.E., H. Preuschoft, and N. Gomberg (eds.). 1979. Environment, Behavior, and Morphology: Dynamic Interactions in Primates. Gustav Fischer, New York.

Ni, X., K.C. Beard, J. Meng, Y. Wang, and D.L. Gebo. 2007. Discovery of the first early Cenozoic Euprimate (Mammalia) from Inner Mongolia. American Museum Novitates 3571:1–11.

Ni, X., D.L. Gebo, M. Dagosto, J. Meng, P. Tafforeau, J.J. Flynn, and K.C. Beard. 2013. The oldest known primate skeleton and early haplorhine evolution. Nature 498:60–64.

Pocock, R.I. 1918. On the external characters of lemurs and Tarsius. Proceedings of the Zoological Society of London, pp. 19–53.

Rose, K.D. 2006. The Beginning of the Age of Mammals. Johns Hopkins University Press, Baltimore.

Rose, K.D., R.S. Rana, A. Sahni, K. Kumar, P. Missiaen, L. Singh, and T. Smith. 2009. Early Eocene primates from Gujarat, India. Journal of Human Evolution 56:366–404.

Seiffert, E.R. 2012. Early primate evolution in Afro-Arabia. Evolutionary Anthropology 21:239–253.

Seiffert, E.R., E.L. Simons, and Y. Attia. 2003. Fossil evidence for an ancient divergence of lorises and galagos. Nature 422:421–424.

Simons, E.L. 1972. Primate Evolution—An Introduction to Man's Place in Nature. Macmillan Press, New York.

Simons, E.L., E.R. Seiffert, P.S. Chatrath, and Y. Attia. 2001. Earliest record of a parapithecid anthropoid from the Jebel Qatrani Formation, Northern Egypt. Folia Primatologica 72:316–331.

Stevens, N.J., P.M. O'Connor, E.M. Roberts, M.D. Schmitz, C. Krause, E. Gorscak, S. Ngasala, T.L. Hierymus, and J. Temu. 2013. Palaeontological evidence for an Oligocene divergence between Old World monkeys and apes. Nature 497:611–614.

Szalay, F.S., and E. Delson. 1979. Evolutionary History of Primates. Academic Press, New York.

Wood Jones, F. 1916. Arboreal Man. Edward Arnold, London.

Zalmoud, I.S., W.J. Sanders, L.M. MacLatchy, G.F. Gunnell, Y.A. Al-Mufarreh, M.A. Ali, A.H. Nasser, A.M. Al-Masari, S.A. Al-Sobhi, A.O. Nahrha, A.H. Matari, J.A. Wilson, and P.D. Gingerich. 2010. New Oligocene primate from Saudi Arabia and the divergence of apes and Old World monkeys. Nature 466:360–363.

Figure 2.1 Geographical distribution of living primates.

The Wet-Nosed Primates

To do primate comparative anatomy, it is helpful if you know the lifestyles of the living primates you wish to understand. This chapter and the following review the behavior and ecology of living primates. All primates, living and extinct, can be subdivided into two major adaptive radiations: the Strepsirhini, or wet-nosed primates, and the Haplorhini, or dry-nosed primates (see table 1.1). Although living primates cluster along the equator today, past geological time periods were warmer and many of the northern continents possessed tropical or subtropical rainforests. Primates roamed the rainforests of central and northern Asia and Europe; even North America had large radiations of tropical primates in the past. These fossil primates have gone extinct or their descendants moved southward toward warmer equatorial climates. Today the Japanese macaques (*Macaca fuscata*) are the most northerly adapted primates (and they contend with snow). Baboons (*Papio ursinus*) and vervets (*Chlorocebus aethiops*) of South Africa living in dry open and woodland environments represent the southern geographic extreme for Africa, being latitudinally equivalent to the most southern-living South American primate, *Cebus apella* (fig. 2.1). The living primates represent at least four taxonomic families on the island of Madagascar, four families in Africa and Asia, and three families in South America. In all, some 66 genera and 348 living species survive at last count. Figure 2.2 displays the living wet-nose primate (lemuriform) taxonomy.

Suborder Strepsirhini
Superfamily Lemuroidea

FAMILY CHEIROGALEIDAE

Mouse and dwarf lemurs (*Allocebus, Cheirogaleus, Microcebus, Mirza,* and *Phaner*; fig. 2.3) are small, arboreal, nocturnal primates that live on the island of Madagascar. Twenty-one species are currently recognized. In terms of body size, many are mouse-sized and all weigh less than 0.5 kg. *Microcebus berthae*, the pygmy mouse lemur, is the smallest living primate, weighing between 24 and 38 g, while *Cheirogaleus major* represents the largest cheirogaleid and weighs 235 to 470 g. Because of their small body size, cheirogaleids are omnivorous, with insects, fruits, and gums making up the largest percentage of food items in their diets (see chapter 6). *Phaner* is a specialized gum feeder, or gumnivore. Cheirogaleids are unusual primates in that they utilize torpor (a form of light hibernation) to conserve energy seasonally. *Microcebus*, and especially *Cheirogaleus*, store fat in their tails for nourishment while hibernating. All cheirogaleids possess large orbits (being nocturnal), a toothcomb, a 2–1–3–3 / 2–1–3–3 dental formula, and shortened limbs. Cheirogaleids are often used as living models for early primate origins due to their small size, primitive social patterns, and generalized locomotor abilities.

Cheirogaleids are generally solitary animals and are described as dispersed within social networks in which adult females often sleep together in nests, in vine and leaf tangles, or in tree holes within the lower canopy during the day. Being solitary means that individuals forage alone and escape from predators on their own. Females raise all infants with no male parental involvement. *Phaner* is unique in being monogamous with female dominance. Mouse and dwarf lemurs, being nocturnal, are olfactory-oriented individuals with frequent scent-marking activities (using fecal, urine, and glandular secretions) within their territories. Scent marking communicates useful information concerning individuality and estrous condition for a species that does not live in a troop. Vocalizations are high-pitched whistles and snorts.

Mouse and dwarf lemurs have a high rate of predation, especially from owls. To counter this high preda-

tion rate, twins are common and females may birth up to four infants per litter, often twice a year. Reproduction is fast paced for cheirogaleids, with gestation periods lasting only two months. Sexual receptivity for females occurs over a one- to two-day time span, and females possess a copulatory plug when not in estrus. To feed their infants, females retain three pairs of nipples. Mothers transport their infants by their mouths early on, but infants develop rapidly and are sexually mature between 8 and 14 months after birth. Cheirogaleids, like *Microcebus*, possess large testes; this feature implies that males engage in sperm competition. Larger testes produce greater sperm volume, giving these males a better chance to father offspring when females mate with more than one male.

Mouse and dwarf lemur locomotion consists of

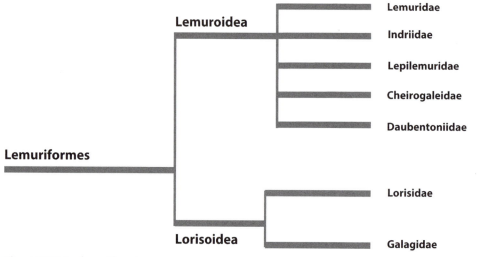

Figure 2.2 Living lemuriform taxonomy.

Figure 2.3 Mouse lemur (*Microcebus murinus, left*) and dwarf lemur (*Cheirogaleus medius, right*). Photos courtesy of David Haring, Duke Lemur Center.

Figure 2.4 Ring-tailed lemur (*Lemur catta, left*) and mongoose lemur (*Eulemur mongoz, right*). Photos courtesy of David Haring, Duke Lemur Center.

arboreal quadrupedalism, leaping, and climbing. They use quadrupedally suspensory movements below branches and are able to cantilever to catch insects from vertical supports (see fig. 8.7). Most retain the common flattened nails of primates, but *Phaner* and *Allocebus* possess pointed or keeled nails compared to those of the other cheirogaleids.

FAMILY LEMURIDAE

Lemurs (*Eulemur, Hapalemur, Lemur,* and *Varecia*; fig. 2.4) represent at least 11 species of diurnal Malagasy primates ranging in size from 600 to 3800 g. All are diurnal, arboreal, frugivore-folivore feeders with home ranges up to 100 ha (one ha = 100 m × 100 m). They all possess long legs. Lemurs are frequent leapers and arboreal quadrupedalists. Lemurs have small, diurnal-sized orbits, a 2–1–3–3 / 2–1–3–3 dental formula, and a toothcomb, and they have lost their lower molar paraconids (see chapter 6).

In contrast to cheirogaleids, lemurs live in cohesive groups. *Eulemur* has a group size of about 10 individuals (range = 5–16) with 3–5 adult males and 2–4 adult females. With the exception of *Lemur catta* (20+ individuals), lemurs live in small groups, being monogamous or polygynous. Lemur troops show female dominance and females remain within their natal group (female philopatry). In *Lemur catta*, matrilines allow females to be the stable core of the troop; reciprocal nursing, kidnapping, and aunting behaviors occur among these related females. This great level of interaction by female kin is evident throughout a *Lemur catta* troop in which females have few close friends among other matrilines. Lemur troops are generally peaceful, with little agonistic behavior outside of the mating season. Social grooming (allogrooming) is common, as are a variety of vocalizations (i.e., alarm calls and cohesion group calls). Male lemurs transfer out of the troop between the ages of 3 and 4.5, thereby entering a new troop. Multimale lemur troops use their "extra" males (non-mating males) as low-cost sentinels. With the exception of the mating season, males play a low-key and insignificant role within a lemur troop.

Lemur life history traits are overall longer than those of other nocturnal primates and their social behaviors are more complex (or monkey-like), as noted above. The lifespan of a lemur is about 20 years compared to the 8-year longevity of a galago. Lemurs have 30- to 40-day estrous cycles and a 4- to 5-month gestation period; first births occur between 2 and 3 years of age. Female lemurs raise their infants without the help of males and generally have just one infant per year. Interbirth intervals for lemurs are around 2 years. Infant mortality rates are high (more than 40%) in lemurs. Larger mothers have better infant survival rates than smaller-sized females.

Oddities

Lemur catta is a semi-terrestrial lemur living in larger troops, composed of 5 to 27 individuals, compared to *Eulemur*. *Lemur macaco* is sexually dichromatic, meaning that males and females have completely different coat colors (i.e., black and brown, respectively, in this case). *Varecia* is a pair-bonded lemur that makes nests and often carries its infants by mouth. Mothers may "park" their young on branches when going off to forage, in contrast to other lemur mothers that carry their infants on their bellies or backs. *Varecia* is also an odd lemur in that its nests are guarded by males, twinning is common, and its alarm call is extremely loud. *Hapalemur* has a specialized diet, eating bamboo laced with cyanide!

FAMILY LEPILEMURIDAE

Lepilemur, the sportive lemur, is another odd Malagasy lemur that is often classified within its own family (Lepilemuridae). The sportive lemur is a nocturnal, arboreal vertical clinger and leaper. It prefers the low canopy and survives well, with high population densities in the drier forests of Madagascar. *Lepilemur* is highly folivorous and this genus has lost its upper incisors while retaining well-developed shearing crests on its molars (see chapter 6). The sportive lemur possesses an enlarged caecum and is considered a hindgut fermenter (see chapter 7). It is known to reingest its own fecal matter (coprophagia), suggesting a tight energy budget for such a small folivore. *Lepilemur* weighs between 600 and 900 g and spends about 80% of its time feeding or resting. Its social organization shows overlapping home ranges for males and females with either male-female or female-female sleeping pairs. Male-male sleeping pairs have never been observed. *Lepilemur* is territorial with a small home range up to 0.5 ha.

FAMILY INDRIIDAE

Indriids (*Avahi, Indri,* and *Propithecus*) are large (800–8000 g) vertical-clinging and leaping Malagasy primates (fig. 2.5). Indriids are socially monogamous, showing female dominance, and all species are highly folivorous. Ten species are currently recognized. In terms of activity, the genus *Avahi* is nocturnal, while *Indri* and *Propithecus* are both diurnal genera. All indriids prefer to feed high in the canopy, eating leaves, flowers, fruit, and bark. They possess large molars with a hypocone, cresty cusps and buccal styles to chop up and shear their leafy diets (see chapter 6). They also possess sacculated stomachs, being foregut fermenters, and long intestines to help digest these leafy materials (see chapter 7). As folivores, they are largely inactive throughout the day within their 4 to 40 ha home ranges.

Indriids live in small groups compared to lemurs. Average group size is about 5, with monogamy being the common social pattern among species within *Indri* and *Avahi*. *Propithecus* tends to have 1–2 adult males or females within multimale groups ranging in size from 3 to 9 individuals. *Propithecus* often has more than one breeding female within this small group. Males and females are social and engage in many allogrooming bouts. Adolescent males and females are forced out of their parents' territory (male and female exogamy).

In terms of life history traits, indriids possess a 40-day estrous cycle, are receptive only for a 2-day

Figure 2.5 Coquerel's sifaka (*Propithecus verreauxi, left*) and the aye-aye (*Daubentonia madagascariensis, right*). Photos courtesy of David Haring, Duke Lemur Center.

timespan per year, and have a 5- to 6-month gestation period. Females start giving birth at 3–4 years of age. One infant is born over a 2-year time interval and infants are weaned at 5–6 months. Life expectancy is between 25 and 30 years.

Oddities

The unique features of indriids include the eerie morning calling bouts as well as a very short tail for *Indri*. *Avahi* is the smallest and the only nocturnal indriid, although it is most active at dawn and at dusk, spending most of the night (65%) resting. Like *Lepilemur*, *Avahi*'s activity pattern is constrained by its

small size and diet. Indriids use throat glands to mark branches as well as urine to demarcate their territory.

FAMILY DAUBENTONIIDAE

Daubentonia, or the aye-aye, may be the strangest of all lemurs inhabiting the island of Madagascar (fig. 2.5). One nocturnal species is recognized, weighing between 2 and 3 kg. The aye-aye is a shaggy-haired, bushy-tailed, bat-eared lemur with pencil-thin middle digits on both hands. *Daubentonia* possesses rodent-like teeth with continuously growing, strong incisors that can open structurally defended food resources (e.g., seeds, hard fruits, and coconuts). In contrast to its robust front teeth, the rest of an aye-aye's dentition is nearly absent (its dental formula is 1–0–1–3 / 1–0–0–3) and only tiny, peg-like molars and premolars remain (see chapter 6). In addition to seeds and fruits, aye-ayes also consume insects. They tap branches, using their thin third finger, to locate embedded insect prey. To capture these wood-boring larvae aye-ayes bite into bark to open cavities. This hole allows their narrow and long third digits to hook and withdraw insect larvae with their claw-like nails.

Aye-ayes are solitary primates behaviorally with male home ranges overlapping those of several females. Home ranges are 6–12 ha in size. In contrast, female home ranges never overlap those of other females. Aye-ayes sleep in well-constructed nests about 17 m high and females park their babies within nests when going off to forage. Aye-aye sleeping trees are intensely marked with urine and bite marks.

Aye-ayes are non-seasonal breeders. They have single births and possess two inguinal nipples, an odd anatomical location across primates. Average gestation period is more than 5 months and infants are sexually mature between 3 and 4 years of age.

Daubentonia prefers the low canopy (0–10 m). Aye-ayes are cautious, arboreal quadrupeds in terms of their locomotor pattern. They can bridge or leap across short gaps, and, like other lemurs, aye-ayes possess hindlimbs that are longer than their forelimbs.

Suborder Strepsirhini
Superfamily Lorisoidea

FAMILY GALAGIDAE

Galagos, or bushbabies, are sub-Saharan African primates with 4 genera (*Euoticus, Galago, Galagoides,* and *Otolemur*) and more than 20 species. They are small, nocturnal primates ranging in size from 60 to more than 1200 g (fig. 2.6). The sexes are similar in body size and canine size. Like cheirogaleids and lorises, galagos are solitary primates but they do possess dispersed social networks or sleeping partners. Galagos forage alone and social groupings, when present, are loose aggregations at best. These dispersed social networks are made up of nesting females that utilize a similar area of the forest and largely share the same home

Figure 2.6 Bushbaby, or galago (*Galago moholi*). Photo courtesy of David Haring, Duke Lemur Center.

range. Female galagos that belong to a different nesting network have home ranges that overlap other networks at the edges. Male galagos tend to sleep alone and have larger home ranges than do females. Their larger home ranges allow males to spatially overlap several female nesting networks, where males monitor estrous conditions. Allogrooming does occur between males and females.

Galagos are known for their mobile ears, large eyes, and long legs, with especially elongated feet. Their active movements tend to stir up insects and galagos pounce on these food items using their hands. Galagos are considered nocturnal visual predators although they also consume gums and fruits. Galagos often utilize a locomotor pattern described as vertical clinging and leaping. The larger galagos, like *Otolemur,* and a few small species are also frequent users of arboreal quadrupedalism. Galago locomotion tends to span a leaping-quadrupedalism gradient.

In terms of life history traits and reproduction, galagos have either two birth peaks per year with twins being common (woodland taxa) or one birth season with one or two infants (rainforest taxa). Galagos are sexually mature between 8 and 12 months of age. An estrous plug forms in the vulva when sexually mature females are not in estrus. The gestation period is 3 months. Mothers carry infants in their mouths when traveling and park infants on branches when going off to forage. Infants are weaned between 2 and 3 months of age. Marking behavior is common among galagos (i.e., urine washing) and these marks are used to identify individuals and to monitor estrus. Life expectancy is 8 years.

FAMILY LORISIDAE

The lorises are nocturnal African (*Arctocebus* and *Perodicticus*) and Asian (*Nycticebus* and *Loris*) primates that resemble little teddy bears since they lack long tails (fig. 2.7). The 6 species weigh between 250 and 1100 g and all are known for their cryptic (hidden) and solitary behaviors. Lorises represent the only Asian strepsirhines alive today. Unlike other strepsirhines, lorises do not leap. They use arboreal quadrupedalism and climbing and have a flexible body and limb anatomy. This body plan makes many of their movements serpentine-like. Lorises have strong grasping hands and feet with widely splayed thumbs and big toes compared to those of other primates. The larger genera (*Perodicticus* and *Nycticebus*) prefer the upper and main forest canopy, while the smaller and more stick-like lorises (*Arctocebus* and *Loris*) prefer the lower levels of the rainforest.

Figure 2.7 A slow loris (*Nycticebus coucang*). Photo courtesy of David Haring, Duke Lemur Center.

In terms of diet, lorises eat fruits, insects, and gums. They tend to feed on insects that contain toxic chemical compounds (i.e., caterpillars, ants, and millipedes). To utilize these poisonous food items, lorises must detoxify them using a slower metabolism and other digestive adaptations. Lorises sneak up on well-camouflaged insects, sniff them out, being olfactory foragers, and grab their prey with their hands.

Male parental care is observed more often among lorises than galagos, but loris mothers spend little time with their mates and do almost all of the infant care themselves. Females exercise dominance in a fairly passive manner toward males; there are few aggressive or agonistic displays. In fact, female lorises are normally quite tolerant of an adult male. Given the solitary habits of lorises, males range farther than females attempting to assess females' estrous status and possible mating outcomes. Lorises have scent glands and both sexes are olfactory in their means of communication.

Given their solitary nature, loris mothers park their infants when going off to forage. The gestation period of lorises, between 5 and 6 months, is longer than galagos. Infants are born once per year and single infants are more common than twins. Infants cling to their mothers when traveling. They are weaned at 4 months. Sexual maturity occurs between 12 and 18 months. In sum, lorises have slower life history traits than their close relatives, the galagos, they eat more toxic food items, and they move in a completely different manner.

Selected References

Alterman, L., G.A. Doyle, and M.K. Izard (eds.). 1995. Creatures of the Dark: The Nocturnal Prosimians. Plenum Press, New York.

Atsalis, S. 2008. A Natural History of the Brown Mouse Lemur. Pearson Prentice Hall, Upper Saddle River, N.J.

Charles-Dominique, P. 1977. Ecology and Behaviour of Nocturnal Primates. Columbia University Press, New York.

Charles-Dominique, P., H.M. Cooper, A. Hladik, C.M. Hladik, E. Pages, G.F. Pariente, A. Petter-Rousseaux, J.J. Petter, and A. Schilling (eds.). 1980. Nocturnal Malagasy Primates. Academic Press, New York.

Charles-Dominique, P., and R.D. Martin. 1972. Behaviour and Ecology of Nocturnal Prosimians—Field Studies in Gabon and Madagascar. Advances in Ethology, Supplement 9, Journal of Comparative Ethology. Verlag Paul Parey, Berlin.

Curtis, D.J., G. Donati, and M.A. Rasmussen (eds.). 2006. Cathemerality. Folia Primatologica 77:1–194.

Falk, D. 2000. Primate Diversity. W.W. Norton and Company, New York.

Fleagle, J.G., C. Janson, and K.E. Reed (eds.). 1999. Primate Communities. Cambridge University Press, Cambridge, UK.

Groves, C. 2001. Primate Taxonomy. Smithsonian Institution Press, Washington, D.C.

Gursky, S., and K.A.I. Nekaris (eds.). 2003. Mating, birthing, and rearing systems of nocturnal prosimians. Folia Primatologica 74:237–376.

Gursky, S., and K.A.I. Nekaris (eds.). 2007. Primate Anti-Predator Strategies. Springer, New York.

Hill, W.C.O. 1953. Strepsirhini Primates: Comparative Anatomy and Taxonomy, Vol. 1. Edinburgh University Press, Edinburgh.

Jolly, A. 1966. Lemur Behavior—A Madagascar Field Study. University of Chicago Press, Chicago.

Kappeler, P.M., and J.U. Ganzhorn (eds.). 1993. Lemur Social Systems and Their Ecological Basis. Plenum Press, New York.

Kappeler, P.M., and M.E. Pereira (eds.). 2003. Primate Life Histories and Socioecology. University of Chicago Press, Chicago.

Lehman, S.M., and J.G. Fleagle (eds.). 2010. Primate Biogeography. Springer, New York.

Martin, R.D. 1990. Primate Origins and Evolution—A Phylogenetic Reconstruction. Princeton University Press, Princeton, N.J.

Mittermeier, R.A., I. Tattersall, W.R. Konstatn, D.M. Meyers, and R.B. Mast (eds.). 1994. Lemurs of Madagascar. Conservation International, Washington, D.C.

Napier, J.R., and P.H. Napier. 1967. A Handbook of Living Primates. Academic Press, London.

Nowak, R.M. 1999. Walker's Primates of the World. Johns Hopkins University Press, Baltimore.

Richard, A.F. 1985. Primates in Nature. W.H. Freeman and Company, New York.

Strier, K.B. 2007. Primate Behavioral Ecology. 3rd edition. Allyn and Bacon, Boston.

Sussman, R.W. 1999. Primate Ecology and Social Structure. Vol. 1: Lorises, Lemurs and Tarsiers. Pearson Custom Publishing, Needham Heights, MA.

Tattersall, I. 1982. The Primates of Madagascar. Columbia University Press, New York.

3

The Dry-Nosed Primates

The dry-nosed, or haplorhine, primates represent a distinct evolutionary radiation relative to the strepsirhine primates (fig. 3.1; see also fig. 1.1). Living haplorhines come in three varieties: tarsiers, monkeys, and apes. The Southeast Asian tarsiers are adaptively similar to African galagos discussed in the previous chapter, while monkeys and apes are quite different from tarsiers or any of the wet-nosed primates.

Suborder Haplorhini
Infraorder Tarsiiformes
Superfamily: Tarsioidea

FAMILY TARSIIDAE

Tarsiers (*Tarsius*; fig. 3.2) live in Southeast Asia and represent at least 5 species weighing between 80 and 150 g. Tarsiers are nocturnal insectivores and faunivores, eating lizards and snakes. They possess a host of unusual anatomical features. They have huge eyes, which are in fact larger than their brains! A tarsier can rotate its head 180 degrees to look backward. Tarsier bodies are highly modified from those of other primates. Tarsier legs are long with especially elongated feet. They also possess long, floppy fingers and toes with broad toe pads, two toilet claws on each foot, and a rat-like tail. They are vertical clingers and leapers that prefer the low canopy, using their large eyes and ears to locate live prey. They typically pounce down

on the ground to capture prey and then leap back to a vertical sapling to feed. Tarsiers possess sharp and pointy incisors, canines, premolars, and tall, pointed molars to crunch prey quickly for consumption.

Tarsiers are active and spectacular leapers that spend most of their time within 2 m of the ground (a near-ground niche). Home ranges vary from 0.5 to 12 ha by species. Tarsiers prefer secondary forest and vertical supports. They sleep in trees, shrubs, ferns, and bamboo thickets, generally just above ground level. Tarsiers often prefer their sleeping sites to be anchored by one large tree. Tarsiers vocalize to maintain their territories and to space themselves out relative to other individuals within the forest. All but *Tarsius spectrum* are solitary species, with male home ranges overlapping female core areas. *T. spectrum* is monogamous. Tarsiers mark with urine and anogenital gland secretions, often rubbing their thighs along branches. Allogrooming is rare.

In terms of life history traits, tarsier mothers typically have one infant per year and are considered non-seasonal breeders. Their gestation period is long for such a small primate at 6 months, in fact, several weeks longer than that of many monkeys, and infants represent 25–30% of their mother's body weight! Tarsiers possess 4–6 nipples although they normally have only a single birth. Females raise their young with no male parental involvement. Babies are parked while mothers leap off to forage. Infants are weaned at 2 months.

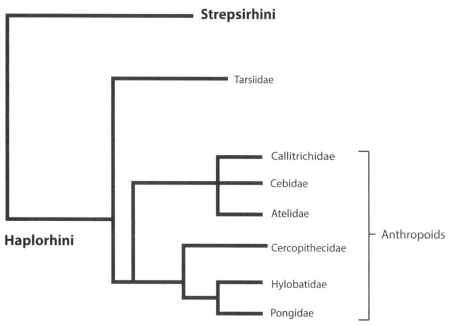

Figure 3.1 Living haplorhine taxonomy.

Figure 3.2 Phillipine tarsier (*Tarsius syrichta*, *left*; photo credit: David Tipling / NPL / Minden Pictures) and spectral tarsier (*Tarsius spectrum*, *right*; photo credit: Ignacio Yufera / FLPA / Minden Pictures).

Figure 3.3 Pygmy marmoset (*Callithrix pygmaea*, *left*; Pete Oxford / Minden Pictures) and common marmoset (*Callithrix jacchus*, *right*; Pete Oxford / Minden Pictures).

Infraorder Simiiformes (or Anthropoidea)
Parvorder Platyrrhini
Superfamily Ceboidea

FAMILY CALLITRICHIDAE

Marmosets and tamarins (*Callibella, Callimico, Callithrix, Cebuella, Leontopithecus, Mico,* and *Saguinus*) are the smallest of the New World monkeys (100–750 g), with females being slightly larger than males (3–10%). Callitrichines have large home ranges, especially given their small size. Callitrichines (fig. 3.3) are quick and active diurnal primates that are cryptic in their behavior. They prefer lowland rainforest and are known for their extensive male parental care, infant care by nonreproductive individuals, and twinning. They possess distinctive claws (or claw-like nails) on all digits but the big toe. Clawed digits are an adaptation for verti

cal support, and callitrichines often cling to large tree trunks to obtain gums and insects. In terms of overall movement patterns, callitrichines are arboreal quadrupedalists that leap, climb, and vertically cling.

Callitrichines are capable large-branch feeders. Some species are opportunistic exudate feeders (i.e., gums, saps, and resins; *Saguinus* sp.), while others (*Saguinus fuscicollis, S. nigricollis, Callimico*) exploit the bark surface for insects. *Leontopithecus* uses manipulative foraging and bark stripping to locate concealed insects. *Callithrix* and *Cebuella* are tree gougers and exudate specialists. Insects, fruits, and gums make up the largest components of callitrichines' diets.

In terms of reproduction, females are sexually mature between 1.5 and 2 years with 4- to 5-month gestation periods. Callitrichines (except *Callimico*) commonly have twins once a year with extensive male parental involvement. The energetic budget of a callitrichine mother is on edge to feed and care for these twins since neonates typically weigh between 10 and 25% of a mother's body weight, in comparison to other platyrrhines where neonates typically weigh < 10% of a mother's weight. The number and large size of neonates taxes callitrichine mothers and this energetic constraint "forces" males to commonly carry their infants 1–2 km per day. Weaning occurs between 30 and 90 days.

The social organization of callitrichines is that of a small group, but its organization is highly variable within and across species. Monogamy, polygyny, and polyandry, as social organizational patterns, occur for different species of callitrichines, and these three social patterns even co-occur within a single species. This variable pattern of social organization shows callitrichines to be flexible in terms of group structure. The environment is believed to play a significant role in determining callitrichine social organization. Usually, one adult female mates with one or two adult males, while other adults within the group are non-breeding helpers. Only one female generally breeds in any social group. In tamarins and marmosets, dominant females are able to suppress ovulation in other adult female subordinates by agonistic interactions. This stress inhibits normal hormone cycles. Callitrichines use urine and glandular secretions to scent-mark for communicating their sexual status.

As small primates, callitrichines are heavily preyed on, usually by raptors, carnivores, and snakes. This intense predation may explain the increase in neonate numbers (i.e., twinning) relative to those of other anthropoid primates, as callitrichines compensate for high mortality with increased reproductive rates. Some tamarins form mixed-species groups, a helpful anti-predator strategy as well.

Oddities

Callimico is an unusual callitrichine in that it has a single birth twice a year, retains its third molars, eats fungus, and is black in coloration in contrast to the more brightly colored callitrichines. *Leontopithecus* and tamarins use holes to sleep in and both have large home ranges. *Cebuella* has twins twice a year, has a tiny home range, and is the most specialized callitrichine, feeding on exudates. *Cebuella* and *Callithrix* both pos-

sess tall and thick incisors, a specialization for gouging holes in trees to produce gum flows.

FAMILY CEBIDAE

Platyrrhine systematics is unresolved at present and the Cebidae represents an unusual mixture of taxa without comprising a simple monophyletic grouping. This chapter groups *Saimiri* (squirrel monkeys), *Cebus* (capuchins), *Aotus* (the owl monkey), and *Callicebus* (the titi monkey) into the single family of Cebidae (see fig. 3.1). These four genera are medium-sized South American primates weighing between 0.7 and 4 kg.

Cebus inhabits all types of forests from Central America to Argentina and is often in mixed-species groups with *Saimiri* (figs. 3.4 and 3.5). *Cebus* and *Saimiri* are arboreal quadrupeds, with *Saimiri* being a more frequent leaper. Both are frugivores that also consume insects. *Saimiri* often scans trees for insects and may feed exclusively on them in certain weeks during the dry season. *Cebus* typically consumes insects while moving between fruit trees and is often a destructive forager, as it breaks branches and strips bark in its search for insects. *Cebus* is also known for its consumption of hard palm nuts. Both *Cebus* and *Saimiri* spend most of their day traveling and foraging. *Saimiri* specializes in highly concentrated food patches like large fig trees, which are dispersed and scattered far apart within rainforests. For example, inter-fruit tree distance at one site was calculated to be 206 m for *Saimiri* compared to 57 m for *Aotus*.

In terms of group size and habitat use, *Saimiri* lives in large (12–100 members), multimale groups. Home ranges vary between 250 and 500 ha with day ranges of 1–4 km. In contrast, *Cebus* species form small (8–12 members), multimale groups, with a male dominance hierarchy and an alpha male. Capuchin males are larger than females. *Cebus* also has large home ranges of 80–300 ha and long 1–4 km day ranges.

Saimiri has one offspring per year and individuals are sexually mature at 2.5–5.5 years of age, although adult size is normally attained at 2 years. Infants are weaned at 6 months. *Saimiri* males fatten to increase size prior to the breeding season to better compete for mates. *Saimiri* species show different relationships between the sexes: egalitarian (*S. oerstedii*), male dominance (*S. sciureus*), and female dominance (*S. bolivien-*

Figure 3.4 Squirrel monkeys (*Saimiri sciureus*; Thomas Marent / Minden Pictures).

sis). Seasonal breeding and birthing are found within *Saimiri*.

Cebus has one offspring every 2 years. They are sexually mature at 5 years of age for females and 7 years for males. Gestation is slightly more than 5 months, with infant weaning occurring between 1 and 2 years. Adult size is achieved between 8 and 10 years. Both *Saimiri* and *Cebus* participate in urine washing and marking activities. *Cebus* is unusual in its prehensile tail use as an adult.

Callicebus is mainly a diurnal frugivore, but different species within this genus prefer either leaves or insects as their most frequently consumed source of protein. Titi monkeys prefer the low canopy and understory, inhabiting dense vegetation in swamps and forests. They prefer small-diameter trees. Titi monkeys utilize a large home range of 4–24 ha and there is little overlap of conspecific home ranges. When approached, their anti-predator response is to freeze or to flee. Their overall preference is to use a cryptic life strategy,

moving slowly through the canopy via arboreal quadrupedalism. *Callicebus* weighs between 1 and 2 kg.

Callicebus has a monogamous social organization with 2–6 individuals per group. Adult males and females are known to entwine their tails when sitting next to each other; they often hold hands and lip-smack. Allogrooming is an important activity among pairs. Both sexes sing dawn duets to maintain their territorial boundaries. Like callitrichines, male titi monkeys are the primary carriers of infants prior to weaning, between 4 and 6 months, with infant transfer to mothers for nursing. Births are seasonal with a single offspring and a gestation period a little more than 4 months. Sexual maturity is between 3 and 4 years. Male and female subadults leave the family unit around 3 years of age.

Aotus is the only nocturnal anthropoid, weighing about 1 kg, and is known as a habitat generalist since it survives in a variety of different environmental conditions and is capable of utilizing alternative food

Figure 3.5 White-faced capuchin monkey (*Cebus capucinus*; Philippe Clement / NPL / Minden Pictures).

resources. *Aotus* has short night ranges and a small home range of 8–10 ha. The home range boundaries of *Aotus* groups overlap extensively. *Aotus* is an arboreal quadruped that prefers to feed in large trees, in contrast to *Callicebus*. *Aotus* is a frugivore that finds food in small, predictable patches. Its sleeping sites are cryptic holes in trees or in dense vegetation. In contrast to *Callicebus*, *Aotus* tends to move noisily through the high canopy at night and when disturbed mobs a potential source of danger.

Owl monkeys are monogamous and males carry infants often. Males also share food with, play with, and groom infants. Single births are the norm for this taxon and weaning occurs at 9 months. Sexual maturity is between 2 and 2.5 years of age and it is common for subadults to leave the family group at 3 years of age.

FAMILY ATELIDAE

The pitheciines (*Cacajao, Chiropotes*, and *Pithecia*; fig. 3.6) are relatively unknown South American monkeys. Pitheciines are 1.5–3.5 kg monkeys that are known for their seed-eating specializations. They all prefer primary rainforest, but *Chiropotes* is found in unflooded and high montane rainforests while *Pithecia* and *Cacajao* utilize seasonally flooded rainforests. *Chiropotes* and *Cacajao* prefer the upper canopy. *Pithecia*, in contrast, utilizes the lower canopy and is an excellent understory leaper. Pitheciines are distinctive for their diet and dental specializations. They have large incisors and canines, small premolars, and small and flat molars, dental features adapted to the consumption of hard seeds and soft fruits. *Chiropotes* often feeds on unripe fruit and seeds. In terms of locomotion *Chiropotes* is an arboreal quadruped with occasional leaping,

Figure 3.6 White-faced saki (*Pithecia pithecia*; Dave Watts / NPL / Minden Pictures).

while *Pithecia* is the most frequent leaper. *Cacajao*, the largest pitheciine with the highest degree of sexual dimorphism, has a short tail and is considered an arboreal quadruped that often uses hindlimb suspension.

The social organization of *Pithecia* is unclear at present. Monogamous family groups were thought to characterize this genus, but larger groups with multimale and multifemale adults have been reported. Further, these small groups often form large aggregations (fission-fusion groups), where social group size and composition vary daily, and individuals within each small unit may move away from each other while foraging. Home ranges may be between 4 and 40 ha, but huge home ranges, up to 200+ ha, have been reported.

Chiropotes lives in flexible multimale groups of up to 30 individuals with fission-fusion distributions while foraging. *Cacajao* also lives in multimale groups with large 500–550 ha home ranges. Larger groups with up to 50 individuals have been noted to split into subgroups of about a dozen individuals. *Chiropotes* and

Cacajao are often seen in mixed-species associations with *Saimiri* and *Cebus*.

Single births are the normal pattern of birthing for pitheciines. Females first give birth at about 3 years of age. Gestation is between 4.5 and 6 months. Mothers provide care to and share food with their infants, which are weaned at 8–12 months.

The atelines (*Alouatta, Ateles, Brachyteles,* and *Lagothrix*; fig. 3.7) are the largest (6–12 kg) of the South American monkeys and all possess muscular and long prehensile tails. Spider monkeys (*Ateles*) and howling monkeys (*Alouatta*) extend their geographic range northward from South America to southern Mexico. *Lagothrix* is Amazonian. In contrast, the woolly spider monkey (*Brachyteles*) has a restricted habitat within the Atlantic coastal forests of Brazil.

For *Alouatta*, home ranges vary between 4 and 20 ha with short 100-m day ranges by groups. This represents a small home range and a short movement pattern for such a large primate. *Alouatta* utilizes dawn howling sessions to space howling monkey groups out

Figure 3.7 Spider monkey (*Ateles*; Mitsuyoshi Tatematsu / Nature Production / Minden Pictures).

across the forest, thereby avoiding contact. Howling monkeys have evolved an enlarged hyoid bone to help these loud calls resonate through the forest.

Social groups within *Alouatta* have 1 or 2 adult males and vary in number between 5 and 20 individuals. Adult males generally are dominant and outrank females. Aggressive behavior can be frequent although overt dominance is rare. Infanticide does occur among howling monkeys. Allogrooming bouts range from rare to common depending on the species. Adult males often move between groups, and both sexes leave their natal birth group.

In terms of reproduction, females first give birth between 4 and 7 years of age. Single infants are born every 2 years with 6- to 7-month gestation periods and no male parental involvement. Aunting behavior occurs among the unrelated adult females and this facilitates infant care. Weaning occurs at 10–14 months.

The diet of howling monkeys is evenly distributed between fruits and leaves. These monkeys possess a large caecum and dental specializations (i.e., large, cresty molars) to process leafy dietary components. Howling monkeys use large time periods of inactivity to digest leaves. They prefer a small number of plant species for bulk but are choosy concerning the leaves they consume. They are arboreal quadrupeds that are able to suspend their bodies below a support using their prehensile tails and their feet, often to obtain a food item.

Ateles (spider monkeys) has a fission-fusion social pattern with 15–40 individuals, in contrast to howling monkeys. These large groups separate into smaller foraging subunits of varying sizes and associations that disperse across the rainforest to seek food. Subgroup sizes correspond to fruit availability. Spider monkeys have been described as multiple central place foragers. They choose several sleeping sites (large trees). They will forage around one sleeping site until its food resources are exhausted and then move to an alternate sleeping site. Subunit sizes adjust to food patches and seasonal variation. Spider monkeys are known for their frequent use of brachiation, arm hanging, and tail-assisted suspensory behaviors while feeding. They also commonly move using arboreal quadrupedalism.

Reproduction occurs throughout the year with females in estrus choosing mates. First births are at 7–8 years of age. The gestation period is 7.5 months. Births are singletons and the interbirth interval is between 25 and 42 months.

Lagothrix, or the woolly monkey, has a social structure similar to that of *Ateles*. Groups range from 4 to 60 individuals, with subgroups dispersing while foraging occurs. Subgroups are not cohesive. Day ranges can cover more than 3 km. *Lagothrix* is a seasonal frugivore-folivore.

Woolly monkey females leave their natal group around 6 years of age and have their first offspring by age 9. Gestation is between 7 and 7.5 months and birth intervals are 3 years.

Although this species can hang by its arms and brachiate with bent elbows, it prefers to move using arboreal quadrupedalism and climbing. *Lagothrix* lacks the forelimb specializations of *Ateles* and *Brachyteles* for frequent brachiation. It prefers the high canopy within rainforests.

Brachyteles arachnoides, or the muriqui, is the largest ateline (12–15 kg) and therefore the largest monkey inhabiting South America. This species lives only in the Atlantic coastal forests of Brazil. Muriquis have a cautious, suspensory locomotor style, especially hanging by their arms and tail. Brachiation with its prehensile tail, quadrupedalism, and climbing movements dominate this species' locomotor repertoire. It has only a vestige of a thumb.

Brachyteles is mostly a folivore but consumes many other food items (i.e., 51% leaves; 32% fruits and seeds; 10% flowers, pollen, nectar, bark, and bamboo). Muriquis have a low metabolic rate; they eat leaves and store these items within their large potbelly to digest. The muriqui is a hindgut fermenter with a large caecum (see chapter 6). They rest 50% of the day and travel only 29%. Muriquis rest at least 6 hours per day with small day ranges varying between 900 and 1400 m depending on the season. They travel more in the summer for fruit. Surprisingly, these species appear to lack parasites and this may be due to eating medicinal plants as part of their diet.

Brachyteles is monomorphic in body size and canine size and thus sexes are co-dominant. The social group is fluid with fission-fusion associations and is described as multimale polygyny. Group size is about 20 with a 1:1 sex ratio. Males remain in their natal troop while females emigrate. This species shows low levels of aggression, usually over resources, not mates. Adult females within a troop will defend food resources from neighboring troops. Males and females spend a considerable amount of time together and are considered to have an egalitarian relationship. Males often stay in close proximity to each other but can travel alone or in groups up to eight in number in search of females in estrus. Male bonding occurs in this species. Affiliative behaviors are common and include embracing, touching, and "chuckling" vocalizations. Females play an active role in choosing males and can avoid unwelcome advances. Females mate with more than one male, and males have very large testes to compete for fatherhood via sperm competition.

Gestation is about 7 months and birth intervals are between 2 and 3 years. Mothers carry babies under their arms, not on their bellies as do most primates. Weaning occurs at 2.5 years with 5- to 6-year-olds going their separate ways.

Infraorder Simiiformes (or Anthropoidea) Parvorder Catarrhini Superfamily Cercopithecoidea

FAMILY CERCOPITHECIDAE

Old World monkeys are a diverse lineage of primates inhabiting Asia, Africa, and the Rock of Gibraltar. This group is composed of two subfamilies: the cercopithecines, or cheek-pouch monkeys, including guenons (*Allenopithecus, Cercopithecus, Chlorocebus, Erythrocebus,* and *Miopithecus*), mangabeys (*Cercocebus* and

Figure 3.8 Sykes monkey (*Cercopithecus albogularis*, *left*; Anup Shah / NPL / Minden Pictures) and black and white colobus monkey (*Colobus guereza*, *right*; Suzi Eszterhas / NPL / Minden Pictures).

Lophocebus), macaques (*Macaca*), and baboons (*Mandrillus*, *Papio*, and *Theropithecus*); and the colobines, or leaf-eating monkeys (*Colobus, Nasalis, Presbytis, Procolobus, Pygathrix, Rhinopithecus, Semnopithecus, Simias*, and *Trachypithecus*; figs. 3.8 and 3.9). All Old World monkeys possess a distinctive molar cusp pattern called bilophodonty (see chapter 6) and all have ischial callosities for attached sitting pads. They range in size from 1 to 50 kg. Some fossil Old World monkeys (e.g., *Theropithecus oswaldi*) weighed in at more than 100 kg! All of these monkeys are diurnal and many are known for their distinctive body markings, natal coats, and mixed-species associations. Cercopithecines are known for their broad incisors, lower molar cusp relief, and cheek pouches in comparison to the colobine monkeys, which have sacculated stomachs and shortened thumbs. Sexual dimorphism occurs among cercopithecids and all adult males are larger in

overall body size and in canine weaponry relative to adult females.

The cercopithecines are mostly African in their geographical distribution with the exception being the Asian macaques. Cercopithecines are more numerous and vary more in appearance than do the colobine monkeys. As a group, they are best viewed as sexually dimorphic, semi-terrestrial, and quadrupedal. Within the cercopithecine radiation, there are several terrestrial species among each of the four living groups (i.e., guenons, mangabeys, baboons, and macaques). In fact, terrestriality has played an important evolutionary role for all of the living Old World monkeys in that this adaptive radiation of monkeys clearly passed through a terrestrial phase early in their evolutionary history, making all of the living arboreal Old World monkeys secondarily arboreal. This scenario contrasts with the evolutionary history of other arboreal lineages of

Figure 3.9 White-handed gibbon (*Hylobates lar, left*; Cyril Ruoso / JH Editorial / Minden Pictures) and siamang (*Hylobates syndactylus, right*; Eric Baccega / NPL / Minden Pictures).

primates and helps to explain the loss of joint mobility in the limbs of all arboreal cercopithecids (see chapters 8 and 9).

Macaques (*Macaca*) are diverse in terms of species numbers (19 are currently recognized) and habitat types. Some live in the rainforests of Asia but most occupy woodlands, swamps, mountains, or grasslands. Macaques are highly terrestrial as a group, but some taxa (e.g., *M. fascicularis*) are more arboreal. All are quadrupedally oriented in terms of foraging and traveling locomotion, being primarily frugivorous, with leaves, bark, flowers, fungi, seeds, and other food items making up the bulk of their far-ranging diets. Home ranges can be quite large (between 100 and 2000 ha).

Macaques live in large multimale-multifemale social groups with female matrilines. Matrilines imply that adolescent males emigrate out of their natal troop. Macaque social groups average between 20 and 30 individuals, but some can be much larger. Adult males are dominant. Females do most of the grooming, often grooming other females and establishing close social bonds. Males are more aggressive than females, although dominance hierarchies exist for both sexes. Female macaques become sexually mature between 3.5 and 6.5 years of age and males a little later. Females demonstrate sexual swellings with birth

intervals of 1.5 years. Alloparenting occurs among mothers. The gestation period is between 5 and 6 months, and single births are the norm. The lifespan of macaques can exceed 30 years.

Savanna baboons (*Papio*) are terrestrial, highly dimorphic, and large (13–25 kg) monkeys. They have long muzzles and tails. Sitting pads are fused in males and separate in females. Grass blades, grass roots, and grass tubers make up the bulk of their diet. Baboons must be alert since they contend with large terrestrial African predators. They are terrestrial quadrupeds during the day, but they sleep in trees at night. Baboons are far-ranging monkeys with low population densities across the grasslands of Africa.

Baboons live in large multimale-multifemale troops that average 40–50 individuals, although some groups include hundreds of monkeys. Troop life is matrilineal with young male emigration with maturity. Male baboons form a dominance hierarchy, and the alpha male with his central hierarchy, the top-ranking males, are the center of the troop. Males are alert, often aggressive, and highly competitive. Savanna baboons have an extensive array of visual signals (e.g., yawns, stares, raised eyebrows, flattened ears, and tail postures) for communication. Grooming is frequently observed and the dominant animals are most often groomed. Friendship building and political alliances are essential social behaviors within a baboon troop since female mate choice is often based on friendships rather than the most dominant male. Females often have male friends that help ensure infant survival. Matrilines also have dominance hierarchies and matrilines compete against each other within a troop. High-ranking females produce healthier and more babies per lifetime than do the more stressed, lower-ranking adult females. Females possess sexual swellings, and baboons lack a birth season.

Mandrills and drills (*Mandrillus*) are frugivorous rainforest baboons. Both are terrestrial and highly dimorphic West African monkeys. Male mandrills are known for their colorful faces, and mandrills and drills both have colorful rear ends with small tails. Mandrills form one-male social groups with 5–10 adult females. Mandrills can form large "herds" numbering in the hundreds. Male mandrills use chest glands to mark their territories. Like baboons, they sleep in trees at night.

Geladas (*Theropithecus*) and hamadryas baboons occupy more arid and often higher-altitude, treeless environments, but are nonetheless behaviorally similar to other baboons. Both geladas and hamadryas baboons use one-male social patterns. Geladas are also known for their naked pink to red chest patches. Among females, this patch becomes redder during estrus and signals sexual receptivity. Hamadryas baboons are known for their mantle of long hair along their shoulders as well as their long facial whiskers. The gelada is a dietary specialist, with grass representing 90% of its diet.

Guenons (*Allenopithecus, Cercopithecus, Chlorocebus, Erythrocebus*, and *Miopithecus*) are smaller and generally arboreal Old World monkeys. They are known for their colorful faces and rear ends, which convey specific communicative signals that coincide with various head bobs, shakes, and stares. Most guenons are frugivorous and quadrupedal above-branch feeders with one-male social groups. Some guenons are terrestrial and some live in multimale-multifemale troops. Guenons represent a highly diverse group in terms of number of species and types of habitats they occupy across Africa.

In contrast to the guenons, mangabeys (*Cercocebus* and *Lophocebus*) are large forest monkeys. They are evenly divided into two terrestrial and two arboreal species. Mangabeys live in multimale-multifemale groups, and males give loud "whoop-gobble" calls to space groups apart.

The colobine monkeys, often called leaf-eating monkeys (fig. 3.8), are generally found in tropical rainforests and most are Asian. They are known for their leaf-eating abilities, as they possess cresty molar teeth (see chapter 6) to shear plant material into small pieces before digestion and a sacculated stomach (see chapter 7). Colobines utilize large salivary glands to help neutralize tannins within these leafy food items and are able to extract water from their folivorous diet. Since leaves are a low-caloric dietary item, copious amounts must be consumed and stored in the stomach, where digestion begins. The leaves are digested by bacteria since primates cannot digest cellulose without bacterial help, making digestion a slow and time-consuming process. As a result, colobines tend to eat large amounts of leafy foods and then rest and digest before moving and foraging again. It is a slow-paced and non-active lifestyle, but their continuing browsing impacts the growth of vegetation.

Besides their folivorous habits, colobines are known for their long legs for leaping and long tails for balance. They have short thumbs and utilize smaller day ranges when compared to cercopithecines.

The hanuman langur (*Semnopithecus entellus*) is one of the most terrestrial of langurs, spending 80% of its day on the ground. Its group size in India varies between 13 and 37 individuals. Hanuman langurs live in multimale (northern troops) or harem (southern troops) troops, with harem troops being more common. These troops range over the home ranges of other troops, making hanuman langurs territorial, with both males and females providing defense of their troop's territory. Within harems, dominant males try to exclude other males at breeding times. Fights resulting in torn ears, facial wounds, or cuts to the tail often occur during this timespan, with more than three-quarters of all adult males showing signs of injury. In comparison, one-third of adult females have similar injuries.

Peripheral males often use investigative efforts called "haunting" to scout out a troop (e.g., gathering information concerning the alpha male or estrous status of females). Defense against haunting males is strictly a male affair. Troop takeover by outside males is commonly observed among one-male harem troops, and invading males are known to commit infanticide to quicken a mother's return to estrus. Older females often help mothers and threaten, chase, and even grapple with invading males in an attempt to protect young infants. To lessen infanticide, mothers attempt counterstrategies that include mating with a new or roving male just before the birth of her infant to confuse fatherhood.

Hanuman females remain in their natal troop for their entire lives and are thus closely related. Matrilines exist within langur troops with a female hierarchy within each matriline. Aunting behavior, or allomothering, is common among colobines. Infants spend half their day with allomothers. In terms of life history traits, colobine mothers have a 15- to 30-month birth interval and a 6.5-month gestation period. A prime female, between 4 and 10 years of age, will give birth every 2–3 years. If there are females in estrus within a troop, there is an increase in visits by nomadic males. The hanuman langur diet is composed of 48% leaves, 45% fruit, and 7% flowers. Tigers are known to prey on hanuman langurs.

In comparison to the hanuman langur, the red colobus monkey (*Piliocolobus badius*) lives in Africa and is able to occupy a wide range of habitats in West and Central Africa. Home ranges vary widely between 9 and 114 ha. Red colobus monkeys live in large multimale-multifemale fission-fusion troops with 12–80 individuals per group with 25–40 individuals representing a typical troop size. Red colobus troops are unusual since they are descended from the male line (patrilineal) with females transferring out of the troop as they mature (female exogamy). Adult males, being relatives, groom each other more often than they groom females. There is a fair amount of male-male competition for mates, however, and females do have a choice of mating partners. Males see their troop as a closed unit and will attack invading males. These troops overlap the home range of other troops, but red colobus monkeys are not territorial over spatial use and are quite tolerant of neighboring groups.

Adult females possess sexual swellings. Gestation periods are close to 6 months. Interbirth intervals are a little more than 2 years. Allomothering in this colobine species is rare since females are unrelated and grooming bouts are rare. They can live into their twenties.

Red colobus monkeys are highly arboreal and prefer to move using quadrupedalism and leaping. They move very little throughout the day about (600 m per day) with travel making up only 9% of their daily activities relative to the time devoted to feeding (45%) or resting (35%). Like other colobines, they eat leaves (73%) and fruit, and generally prefer higher-quality foods than the food items consumed by sympatric black and white colobus monkeys (i.e., a higher percentage of young leaves, flowers, buds, and fruit).

Red colobus monkeys are often preyed on by crowned-hawk eagles and chimpanzees. These monkeys possess a complex set of vocalizations, including alarm calls for specific predators (i.e., pythons, eagles, and crocodiles).

Infraorder Simiiformes (or Anthropoidea) Parvorder Catarrhini Superfamily Hominoidea

FAMILY HYLOBATIDAE

Gibbons are the smallest (5–11 kg) and taxonomically the most diverse of all the living apes (fig. 3.9). At least nine species of gibbons are recognized throughout Southeast Asia (i.e., southern China, Thailand, Burma, Borneo, Sumatra, and Java). Gibbons are characterized by their long arms for brachiation and arm-hanging postures as well as their lack of sexual dimorphism. Gibbons possess the longest forelimbs of all primates, as well as long, hook-like fingers, and the odd anatomical feature called ischial callosities, commonly found among cercopithecids. Gibbons prefer to sit rather than to lie down compared to the other living apes. They do not make night nests like those constructed by the great apes.

The long arm span of gibbons allows them to be extremely efficient as terminal-branch feeders. Gibbons prefer ripe fruit, especially figs, and must travel long distances to find these rich food patches. Their day range equals 1–2 km of movements. Given their preference for rich but widely spaced food patches, gibbon home ranges are large (30–50 ha). In contrast, siamangs, a gibbon species that consumes more leaves, has smaller home (23 ha) and day (1 km) ranges.

Gibbons prefer moist primary rainforest, inhabiting the mid- to upper canopy levels of the forest during the day. A typical day is divided equally among resting, feeding, and traveling, with a small percentage of the day devoted to calling. In terms of social structure, gibbons are monogamous with up to four offspring per family group. Extra-pair matings occur with neighboring gibbons 12% of the time, however. Thus, males are more attentive to females than females are to them and males prefer to guard their mates rather than to pursue extra-pair matings. Either sex may initiate mating, although females lack sexual swellings, normally a visual signal for males.

In terms of life history traits, single births occur every 4 years for gibbons with gestation lasting about 7 months. Gibbons are sexually mature at 6–8 years of age and live to more than 30 years. Young adults of both sexes emigrate from their natal group at around 10 years of age. Siamang adult males carry and care for their young more often than other male gibbons.

Gibbons are territorial and will defend their core areas. Although fighting is a rare event, both males and females will participate in territorial defense. Daily calling or singing bouts, including dueting, help space gibbon groups, thereby minimizing agonistic encounters. These songs are species-specific with siamangs possessing an unusual anatomical structure, an inflatable air or throat sac that inflates like a balloon and functions as a resonance chamber when siamangs call.

FAMILY PONGIDAE

There are four species of great apes. One species, the orangutan (*Pongo pygmaeus*), is Asian, while the other three (*Gorilla gorilla*, *Pan troglodytes*, and *Pan paniscus*) are African. Of all of the living primates, the great apes are the most closely related to humans. In fact, humans and chimpanzees are genetically more closely related to each other than chimpanzees are to gorillas. Great apes are also distinguished from other living primates in their superior cognitive abilities. They are self-aware, great social manipulators, tool makers and users, and excellent problem solvers.

Orangutans (*Pongo*) currently live only on the islands of Borneo and Sumatra (northern tip only—

tigers are still present in these Sumatran forests), although orangutans formerly inhabited mainland Asia. Orangutans represent the largest arboreal mammal. Orangutans have unusual faces with small, close-set eyes (fig. 3.10). Adult males are particularly known for their large body sizes relative to those of adult females, with males being twice their size; they possess large canines, cheek-pads, beards, and large laryngeal sacs or throat pouches. Males can hang their entire 150-plus-kg bodies below a branch using only 2 fingers and do pull-ups! Orangutans possess a 2.4-m arm span.

Orangutans prefer lowland forests close to rivers. They travel little each day, with only 50–500-m day ranges. Their movement pattern is often described as cautious and climbing-oriented. They can brachiate slowly, climb upward or down, or fist-walk when on the ground (occasionally), but they generally move through the canopy with climbing and suspensory body postures and are described as arboreal clamberers. Half of an orangutan's day involves feeding, roughly 40% of the day is spent resting, and the remaining hours of the day (11%) are for traveling.

Orangutans are selective feeders, like gibbons, but they need an enormous amount of food daily due to their large body size. They prefer fruit but will consume bark, leaves, piths, and insects. Orangutans have been observed to eat lorises on occasion. They consume 400 types of food; this high number of food items requires good memories as to the types of edible foods and their locations. Orangutans in Sumatra, but not Borneo, have been observed to use sticks as tools to rub off irritant hairs on fruits, to fish for termites, and to extract honey. Males, being larger, generally feed longer each day than do females.

The rainforests in Asia differ from those in Africa in two important ways. First, the trees are smaller in their crown diameter and thus possess smaller fruit patches. Second, these dipterocarp forests have an unusual characteristic called mast fruiting, where up to 90% of rainforest trees produce fruit during the same time of the same year. Unfortunately for frugivores, mast fruitings do not occur every year but at 2- to 8-year intervals. This pattern of synchronous fruiting within Asian forests oversupplies the fruit consumers, allowing a large number of seeds to disperse. This mast-fruiting strategy by trees means that in most

Figure 3.10 Mountain gorilla (*Gorilla gorilla*, *left*; Mitsuyoshi Tatematsu / Nature Production / Minden Pictures) and an orangutan (*Pongo pygmaeus*, *right*; Cyril Ruoso / JH Editorial / Minden Pictures).

years fruit is scarce in the forests of Southeast Asia, a problem for orangutans and other Asian frugivores like gibbons.

There is strong male-male competition for mates in orangutans and they are highly sexually dimorphic. Male orangutans come in two varieties: alpha, or resident, males and non-flanged, or wandering, males. These two varieties are the result of different hormone profiles in males (male bimaturism). Alpha males have undergone all of the adolescent hormonal changes to look and act like a mature adult male. In contrast, the males without cheek-pads lack these specific hormonal triggers and have suppressed reproduction. This condition can last up to 30 years in non-flanged males. Non-flanged males tend to be more social and actively seek and follow potentially fertile females (i.e., females without infants).

Male residents can wander away from their core areas, but adult male orangutans are intolerant of other males and their long call helps to separate them. These long calls are territorial calling bouts to space males

apart and to attract females to their spatial residency. Adult male orangutans either avoid each other or fight. Cheek-padders do not play.

Orangutan adult females lack sexual swellings, and they clearly prefer resident males. Orangutans are solitary primates, however, and roving male promiscuity helps to explain their odd solitary social pattern since the amount of food within an Asian forest is relatively small and widely dispersed.

The only real cohesive "group" in orangutans is the mother and her infant. Adult females spend about 90% of their lives alone with their infants. Females will socialize with other females and their young, especially in large fruiting trees when superabundant fruit crops are available. Temporary associations of up to eight individuals have been reported. Female home ranges overlap, but there is no active female bonding and friends only rarely embrace, kiss, hug, hold hands, or groom each other.

The home range of a resident male overlaps that of several adult females, a spatial arrangement similar to that of male galagos. The home range of orangutans is generally 1.5 km for both male and female orangutans. The size of the home range and the subsequent space between individual orangutans prevent resident males from defending females or possessing exclusive access to fertile females. Adult females remain close to resident males for their "umbrella-like" long-call protection that acts to keep other male orangutans away from this area, a strategy that helps prevent infanticide by roving males. Resident males therefore act as beacons for mothers. Females will form consortships with specific males, and they express strong mating preferences for and against certain males. Females often initiate contact and consortships with males and terminate their time together.

A female orangutan has her first infant at 15 years of age, and mothers will nurse infants until they are 8. Orangutan mothers invest a substantial amount of time (4–8 years) in rearing each infant, and they only birth 3 babies per lifetime, although they live into their fifties and sixties. The birth interval of orangutans is the longest of all primates at 6–7 years. Males invest no time in infant care. Orangutans make night nests.

Gorillas (*Gorilla*) are found in only two areas of tropical Africa: western central Africa (Cameroon, Nigeria, Gabon, Equatorial Guinea, and the Democratic Republic of the Congo) and the eastern edge of the Congo Basin (eastern part of the Democratic Republic of the Congo [formerly Zaire], Uganda, and Rwanda). Gorillas occupy forests from sea level up to 4 km in elevation. Mountain gorillas are restricted to montane and bamboo forests above 2.7 km (fig. 3.10). Gorillas are the largest living primates, with adult males being more than twice the size of adult females. Adult females can weigh up to 115 kg while males can exceed 205 kg. At this size, gorillas have few predators although leopards will occasionally prey on young gorillas.

Gorillas are adept in trees but have secondarily adapted to life on the ground. They are a largely terrestrial species, using terrestrial quadrupedalism (often described as knuckle walking) as their means of traveling between feeding and resting sites.

Gorillas are folivores (85% of their diet), with fruit, bark, and insects making up the rest of their diet. Due to their large body size, gorillas spend 6–8 hours per day feeding. They have never been observed to use tools or to eat meat. The day range in mountain gorillas is small, usually between 0.5 and 1 km, while the western lowland gorilla travels more widely and eats a larger percentage of fruit. Gorillas are dexterous feeders in that they are able to consume nettles and other structurally defended food items by using precise and careful hand and finger movements to strip troublesome areas away.

Gorilla life is a low-key affair for the most part. They spent most of the day resting (40% of the day), feeding (30%), and travel-feeding (30%). Their food items literally surround them as they travel through the forest. Like all folivores, they need to rest to digest. Gorillas have been described as primates living in a great big salad bowl. They have large stomachs as storage chambers for an overall low-quality diet. Large male gorillas consume 60 kg of food daily.

Gorillas live in harem groups with usually a single dominant male. The silverback male, the breeding male, is protective of the group, recruits females, and is highly aggressive toward strange outside males. Silverback males will tolerate an adult son within the harem, generally when he is older, and thus gorilla groups (albeit larger ones) can have multiple breeding males for a time. Silverbacks hold the harem, normally consisting of 3–4 sexually mature females and their young, together. Females are unrelated and their status in the group is determined by the order of acquisition. The first female to bond with a silverback male will always hold the first rank, followed by the second acquired female, and so on, regardless of female size or other social attributes. Harems average 10 individuals but have been observed to be as large as 30 individuals. The silverback male is the dominant animal and his large body size, skull size, and canines indicate intense sexual selection due to male-male competition for mates. The silverback is groomed most often since he has the highest status, and he rarely grooms other gorillas.

Although gorilla groups are usually peaceful and generally avoid adult clashes, adult males will fight and contest for their own harems. Silverback males commonly possess broken canines or head wounds suffered from these intense fights. Silverback males in harems can sire two to three times as many offspring than a male with a single female and thus there is a tremendous amount of fitness to be gained by win-

Figure 3.11 Chimpanzee mother (*Pan troglodytes*; Anup Shah / NPL / Minden Pictures).

ning a harem. Male breeding tenure in a harem is generally long and can exceed 16 years. If the silverback dies or is defeated, the harem will break up. Although lone male gorillas are common in African forests, lone females, with or without infants, usually enter or are recruited to enter a harem rather quickly. Gorilla home ranges overlap those of other gorilla groups, making it difficult for females to avoid silverbacks for long. Mothers try to defend their infants against these males and many take wounds in active defense of their infants, but given the size and strength differentials males prevail. Male gorillas will commit infanticide if the infants are not weaned, bringing females back into estrus more quickly, and these mothers will enter the harem and subsequently breed with the infanticidal male. Female mate choice is largely judged on the criterion of male protective abilities.

Adult female gorillas have their first babies around 10–11 years of age. Males begin fatherhood later, at around 15 years of age. Female gorillas do not have sexual swellings. Gestation is 8.5 months; mothers tend to produce only 4 or 5 babies per lifetime with birth intervals of 3.8 years. Infant mortality is highest in the first 3 years of life, with about one-third of those deaths attributed to infanticide. Gorilla infants sleep in their mothers' nests until they are 3–4 years old. Mountain gorillas generally make night nests on the ground, while the western lowland gorillas make about one-third of their nests in trees. Adolescent males and females emigrate away from the group. Silverback males are protective but generally provide little parental investment. Gorillas live well into their fifties.

Chimpanzees (*Pan troglodytes*; fig. 3.11) are restricted to rainforests in Central Africa and savanna-woodland environments from West Africa (from Senegal, Gambia, and Sierra Leone) to the western edge of Tanzania in East Africa. Chimpanzees are sexually dimorphic, with males being 25% larger in body size relative to females, but chimpanzees are far less extreme in their dimorphism relative to gorillas or orangutans. Chimpanzee males' average body weight is 60 kg while females weigh in at 47 kg. They have no formal predators but are wary of leopards.

Chimpanzees are frugivores, with 75% of their annual diet being fruit. Chimpanzees also consume leaves, including medicinal plants when ill, blossoms, seeds, bark, insects, and meat. They have been noted to eat 149 different types of plants.

Like gorillas, chimpanzees are adapted for an arboreal and a terrestrial lifestyle. They are adept at climbing into trees and will forage and build night nests in the canopy. They travel on the ground using terrestrial quadrupedal knuckle walking about 85% of all locomotor bouts. They move over a large daily distance, especially males, which average 4.9 km per day relative to females that move 3 km per day.

Chimpanzees live in large multimale-multifemale communities numbering up to 80 individuals. They form fluid fission-fusion associations such that the community is never completely together at one time. They are known for their lifelong family ties. Females leave their natal group while males remain, making males within a chimpanzee community kin. This kinship promotes strong male-male bonds, the complete opposite of the gorilla or orangutan male adaptive strategies. Adult male chimpanzees often patrol the boundaries of the community. Male chimpanzees are dominant to females, and males often engage in power plays and coalition formations resembling triangular-based political maneuvers. Each community is closed to outside males. Chimpanzee males, within a community, share females and thus reproductive potential. The male role in the community is largely one of protector of the community's territory. Male chimpanzees have been reported to invade other communities and to kill outside males.

Females often forage alone and do not display an obvious dominance hierarchy like that of males. In some forests, female chimpanzees have greater social positions and often build strong female friendships. These alliances take a more active role in male conflicts. High-ranking females tend to have better access to food resources, they live longer, their infants survive better, they produce more infants per lifetime, and their daughters mature earlier than those of low-ranking females. In contrast, low-ranking females show a higher mortality rate, as well as a shorter overall lifespan, and their daughters produce sexual swellings much later in life. Female competition for food can be intense, especially between resident and immigrant females. Female chimpanzees are the ecological sex in that their reproductive gains are intimately tied to the environment. Females are also choosy concerning potential mating partners.

Females are sexually mature at 7 and they have their first infant at 14 years of age. Males mature later at age 12. Females are generally less social than males although mother-infant bonds are strong. Gestation in chimpanzees is about 8 months. Birth intervals are between 3.5 and 5 years, with an adult female producing 4 offspring per lifetime. Usually, a chimpanzee community has only one or two infants annually. Infant mortality is 30% in the first year of life. Weaning occurs in the fourth year. Mothers help their sons dominate females. Like the other great apes, chimpanzees have a low reproductive rate and thus they must be intelligent and competitive to survive in their environment. The mean lifetime for chimpanzees is less than 15 years, although some have lived into their forties in the wild.

Chimpanzees are known for their exuberant and pronounced sexual behavior. Females in estrus prefer males that groom them, share food, and spend time with them. Although alpha males can force females to accompany them for short periods of time, consortships with a mutually agreeable couple tend to produce a greater number of conceptions. Consortships average between 9 and 10 days but can last more than a month. In contrast, dominant males will allow subordinate males copulations with females but these usually occur with non-fertile females. Immigrant females are far more active and less choosy when it comes to sexual activity than are resident females. Genetic evidence reveals, however, that perhaps as many as half of the infants fathered within a chimpanzee community are fathered by outside males! Female chimpanzees are showy when coming into estrus since they produce large, pink-colored sexual swellings (labial swellings), an obvious visual signal to males. Females often mate when they cannot conceive, with a 35-day menstrual cycle; sexual swellings occur about half of the cycling period.

When adolescent females transfer out of their natal

community, they are met with stiff resistance from resident females in the new community. Several visits with longer and longer stays eventually allow them entry into the new community. Immigrant females are often seen with the adult males of the new community, thereby gaining immunity from resident females, which act to defend their core resource areas. High-ranking females are known to snatch and kill infants from lower-ranking females.

Chimpanzees often use tools in the wild. Males tend to use weaponry (e.g., clubs, flails, and missiles) eight times more often than females, while females more often use tools for foraging. Termite fishing, the obtaining and eating of termites within termite nests using stiff but flexible sticks as fishing rods, is performed more often and for a longer duration by females than males. This is also the case for ant dipping and nut cracking, female gathering techniques for a localized food resource. In contrast, chimpanzee hunting, often for red colobus monkeys, is a male-dominated activity (36 times more often) that occurs more frequently when females in estrus or females with sexual swellings are in the party. These cooperative hunts are quite sophisticated with the hunter, not the alpha male, being able to redistribute the meat, usually to females for sex. Hunting success is tied to the size of the hunting party.

Bonobos (*Pan paniscus*) are restricted to rainforest south of the Zaire River in Central Africa in the Democratic Republic of the Congo, formerly called Zaire. Bonobos are more slender than chimpanzees, with long, well-muscled limbs, a black face with hair tufts and black ears, a white tail tuff, and a relatively smaller head than chimpanzees. Bonobos' average body weights are 43 kg for males and 37 kg for females. These body weights are comparable to the size of eastern chimpanzees (*Pan troglodytes schweinfurthii*). Bonobos are more arboreal than chimpanzees but still use terrestrial knuckle walking to travel. Bonobos eat a greater percentage of terrestrial herbaceous vegetation (i.e., pith—a more gorilla-like dietary resource) than chimpanzees. Fruit patches determine the size of the foraging party. Bonobos eat insects and are reported to eat duikers.

There are no male bands among bonobos and traveling parties are mixed sex. Male kinship is focused on a male's mother rather than on other males. Females often dominate males. A bonobo community can be viewed as a more female-centered community than a chimpanzee community. Community size varies between 10 and 58 individuals. The male to female sex ratio is 1:1 in bonobos in contrast to the 4:1 ratio in chimpanzees. Female bonding and male aversion are more common among bonobos, the opposite pattern seen in chimpanzees.

Sexual activity is much more common among bonobos, with sexual swellings representing 75% of a sexual cycle. Sexual swellings last more than 20 days in bonobos in comparison to only 9.5 days in chimpanzees. Bonobos resume sexual swellings one year after a birth. Most sexual encounters have nothing to do with reproduction but instead function as social modifiers. Sexual encounters keep rivalries low. Females beg for food and negotiate with copulations (i.e., food for sex). Females are social, grooming each other and performing homosexual activities with each other. Homosexual hunching or genital-genital rubbing often occurs among bonobos, primarily between females. Sex represents a social function of reducing tension during close feeding periods, thereby ensuring a more peaceful co-existence. Male bonobos are less sexually competitive than male chimpanzees and rarely recruit supporters. Physical violence is rare and male bonobos never make alliances. The long-lasting sexual swellings and frequent copulations by females tend to obscure fatherhood among the males. Males are simply attracted to these swellings and many believe the whole purpose of this extended sexual receptivity in bonobos is to prevent infanticide by obscuring fatherhood. Female bonobos are sexually receptive about 50% of their life compared to about 5% for chimpanzees. Copulations increase with male rank.

Although sexual activity is high among bonobos, there is no corresponding increase in the number of births. Females rejoin their group immediately after infants are born. In contrast, female chimpanzees stay away from groups after giving birth. Nursing in bonobos occurs until infants are 5 years old and birth intervals are also about 5 years. Sexual swellings in females develop at 7 years of age when females leave their natal group. These swellings reach full size at age 10 and first offspring arrive when a female is between 13 and 15 years of age. Bonobos are known to live into their fifties. Tool use or group hunts have never been observed in bonobos to date. On the other hand, as bonobos are studied over longer periods of time, what had been thought to be significant differences between chimpanzees and bonobos seem to be blurring.

Selected References

Altman, J. 1980. Baboon Mothers and Infants. Harvard University Press, Cambridge, MA.

Altmann, S.A., and J. Altmann. 1970. Baboon Ecology—African Field Research. University of Chicago Press, Chicago.

Boehm, C. 1999. Hierarchy in the Forest—The Evolution of Egalitarian Behavior. Harvard University Press, Cambridge, MA.

Boesch, C., G. Hohmann, and L. Marchant (eds.). 2002. Behavioural Diversity in Chimpanzees and Bonobos. Cambridge University Press, Cambridge, UK.

Byrne, R. 1995. The Thinking Ape—Evolutionary Origins of Intelligence. Oxford University Press, Oxford, UK.

Caldecott, J., and L. Miles (eds.). 2005. World Atlas of Great Apes and Their Conservation. University of California Press, Berkeley.

Campbell, C.J., A. Fuentes, K.C. Mackinnon, M. Panger, and S.K. Bearder (eds.). 2007. Primates in Perspective. Oxford University Press, Oxford, UK.

Davies, A.G., and J.F. Oates (eds.). 1994. Colobine Monkeys—Their Ecology, Behavior and Evolution. Cambridge University Press, Cambridge, UK.

de Waal, F. 2007. Chimpanzee Politics—Power and Sex among Apes. Johns Hopkins University Press, Baltimore.

Dixson, A.F. 1981. The Natural History of the Gorilla. Columbia University Press, New York.

———. 1998. Primate Sexuality: Comparative Studies of the Prosimians, Monkeys, Apes, and Human Beings. Oxford University Press, Oxford, UK.

Dunbar, R.I. 1988. Primate Social Systems. Comstock Publishing Associates, Ithaca, N.Y.

Falk, D. 2000. Primate Diversity. W.W. Norton and Company, New York.

Fleagle, J.G. 1999. Primate Adaptation and Evolution. Academic Press, New York.

Fleagle, J.G., C. Janson, and K.E. Reed (eds.). 1999. Primate Communities. Cambridge University Press, Cambridge, UK.

Ford, S.M., L.M. Porter, and L.C. Davis (eds.). 2009. The Smallest Anthropoids. Springer, New York.

Fossey, D. 1983. Gorillas in the Mist. Houghton Mifflin Company, Boston.

Galdikas, M.F., N.E. Briggs, L.K. Sheeran, G.L. Shapiro, and J. Goodall (eds.). 2001. All Apes Great and Small. Vol. 1: African Apes. Kluwer Academic–Plenum Publishers, New York.

Garber, P.A., A. Estrada, J.C. Bicc-Marques, E.W. Heymann, and K.B. Strier. 2009. South American Primates. Springer, New York.

Gautier-Hion, A., F. Bourliere, J.P. Gautier, and J. Kingdon (eds.). 1988. A Primate Radiation—Evolutionary Biology of the African Guenons. Cambridge University Press, Cambridge, UK.

Glenn, M.E., and M. Cords (eds.). 2002. The Guenons—Diversity and Adaptation in African Monkeys. Kluwer Academic–Plenum Publishing, New York.

Goodall, J. 1971. In the Shadow of Man. Rev. edition. Houghton Mifflin Company, Boston.

———. 1986. The Chimpanzees of Gombe—Patterns of Behavior. Belknap Press of Harvard University Press, Cambridge, MA.

Groves, C. 2001. Primate Taxonomy. Smithsonian Institution Press, Washington, D.C.

Gursky, S.L. 2007. The Spectral Tarsier. Pearson Prentice Hall, Upper Saddle River, N.J.

Harcourt, A.H., and F.M. de Waal (eds.). 1992. Coalitions and Alliances in Humans and Other Animals. Oxford Science Publications, Oxford, UK.

Hrdy, S.B. 1977. The Langurs of Abu—Female and Male Strategies of Reproduction. Harvard University Press, Cambridge, MA.

Kano, T. 1992. The Last Ape—Pygmy Chimpanzee Behavior and Ecology. Stanford University Press, Stanford, Calif.

Kappeler, P.M. (ed.). 2000. Primate Males—Causes and Consequences of Variation in Group Composition. Cambridge University Press, Cambridge, UK.

Kappeler, P.M., and M.E. Pereira (eds.). 2003. Primate Life Histories and Socioecology. University of Chicago Press, Chicago.

Kingdon, J. 1997. The Kingdon Field Guide to African Mammals. Academic Press, New York.

Kinzey, W.G. (ed.). 1997. New World Primates—Ecology, Evolution, and Behavior. Aldine de Gruyter, New York.

———. 1987. The Evolution of Human Behavior: Primate Models. State University of New York Press, Albany.

Kummer, H. 1968. Social Organization of Hamadryas Baboons—A Field Study. University of Chicago Press, Chicago.

Martin, R.D. 1990. Primate Origins and Evolution—A Phylogenetic Reconstruction. Princeton University Press, Princeton, N.J.

McGrew, M.C. 1992. Chimpanzee Material Culture—Implications for Human Evolution. Cambridge University Press, Cambridge, UK.

McGrew, W.C., L.F. Marchant, and T. Nishida (eds.). 1996. Great Ape Societies. Cambridge University Press, Cambridge, UK.

Milton, K. 1980. The Foraging Strategy of Howler Monkeys—A Study in Primate Economics. Columbia University Press, New York.

Mittermeier, R.A., A.B. Rylands, A. Coimbra-Filho, and G.A.B. Fonseca (eds.). 1988. Ecology and Behavior of Neotropical Primates. Vol. 2. World Wildlife Fund, Washington, D.C.

Napier, J.R., and P.H. Napier. 1967. A Handbook of Living Primates. Academic Press, London.

———. 1985. A Natural History of the Primates. MIT Press, Cambridge, MA.

Norconk, M.A., A.L. Rosenberger, and P.A. Garber (eds.). Adaptive Radiations of Neotropical Primates. Plenum Press, New York.

Nowak, R.M. 1999. Walker's Primates of the World. Johns Hopkins University Press, Baltimore.

Porter, L.M. 2007. The Behavioral Ecology of Callimicos and Tamarins in Northwestern Bolivia. Pearson Prentice Hall, Upper Saddle River, N.J.

Richard, A.F. 1985. Primates in Nature. W.H. Freeman and Company, New York.

Robbins, M.M., P. Sicotte, and K.J. Steward (eds.). 2001. Mountain Gorillas—Three Decades of Research at Kariosoke. Cambridge University Press, Cambridge, UK.

Rodman, P.S., and J.H. Cant (eds.). 1984. Adaptations for Foraging in Nonhuman Primates. Columbia University Press, New York.

Rowe, N. 1996. The Pictorial Guide to the Living Primates. Pogonias Press, Charlestown, R.I.

Russon, A.E. 2000. Orangutans—Wizards of the Rain Forest. Firefly Books, Buffalo, N.Y.

Russon, A.E., K.A. Bard, and S.T. Parker (eds.) 1996. Reaching into Thought—The Minds of the Great Apes. Cambridge University Press, Cambridge, UK.

Rylands, A.B. (ed.) 1993. Marmosets and Tamarins—Systematics, Behavior and Ecology. Oxford University Press, Oxford, UK.

Schaller, G.B. 1963. The Mountain Gorilla—Ecology and Behavior. University of Chicago Press, Chicago.

Smuts, B.B., D.L. Cheney, R.M. Seyfarth, R.W. Wrangham, and T.T. Struhsaker (eds.). 1986. Primate Societies. University of Chicago Press, Chicago.

Stanford, C.B. 1998. Chimpanzee and Red Colobus—The Ecology of Predator and Prey. Harvard University Press, Cambridge, MA.

Strier, K.B. 1992. Faces in the Forest—The Endangered Muriqui Monkeys of Brazil. Oxford University Press, Oxford, UK.

———. 2007. Primate Behavioral Ecology. 3rd edition. Allyn and Bacon, Boston.

Struhsaker, T.T. 1975. The Red Colobus Monkey. University of Chicago Press, Chicago.

Strum, S.C. 1987. Almost Human—A Journey into the World of Baboons. Random House, New York.

Sussman, R.W. 1979. Primate Ecology: Problem Oriented Field Studies. John Wiley and Sons, New York.

———. (ed.). 2000. Primate Ecology and Social Structure. Vol. 2: New World Monkeys. Pearson Custom Publishing, Needham Heights, MA.

Taylor, A.B., and M.L. Goldsmith (eds.). 2003. Gorilla Biology—A Multidisciplinary Perspective. Cambridge University Press, Cambridge, UK.

Terborgh, J. 1983. Five New World Primates—A Study in Comparative Ecology. Princeton University Press, Princeton, N.J.

Tuttle, R.H. 1986. Apes of the World—Their Social Behavior, Communication, Mentality and Ecology. Noyes Publications, Park Ridge, N.J.

Van Schaik, C. 2004. Among Orangutans—Red Apes and the Rise of Human Culture. The Belknap Press of Harvard University Press, Cambridge, MA.

Whiten, A,. and M. Widdowson (eds.). 1992. Foraging Strategies and Natural Diet of Monkeys, Apes and Humans. Oxford Science Publications, Oxford, UK.

Wrangham, R.W., W.C. McGrew, F.B.M. de Waal, and P.G. Heltne (eds.). 1994. Chimpanzee Cultures. Harvard University Press, Cambridge, MA.

Wright, P.C., E.L. Simons, and S. Gursky. 2003. Tarsiers—Past, Present and Future. Rutgers University Press, New Brunswick, N.J.

4 Musculoskeletal System

Before we launch into the comparative anatomy of all living primates, it is worthwhile to consider bone biology, the muscles that attach to bones, and the mechanical abilities the musculoskeletal system allows. Bones make up the skeletal structure of a vertebrate's body and as such they (1) support the body's soft tissues (e.g., as rigid attachment points for muscles); (2) protect vital organs (by enclosing the brain or spinal cord, for instance); and (3) have a physiological role in the production of red blood cells, the storage of fat tissue, and the body's monitoring of minerals like calcium and phosphate. Without bones, a primate body would simply lack structure and be completely dysfunctional.

BONE BIOLOGY

Bone is a specific type of biological tissue that is unique to the body in that bone is composed of both an organic and an inorganic component. The organic component represents about one-third of the overall composition of any bone in a body. Collagen, a stretchy protein also found in cartilage, tendons, ligaments, and skin, makes up 90% of the organic content of bone, providing elasticity, flexibility, and strength. Hydroxyapatite, a type of calcium phosphate, is the inorganic, or mineral, component in bone tissue. Hydroxyapatite crystals impregnate the matrix of collagen fibers, providing stiffness to bone tissue. The amount of mineralization, and of course bone strength, varies depending on the amount of colla-

gen relative to hydroxyapatite. Strong bones utilize a higher percentage of collagen, while stiff bones have more calcium phosphate. Professor John D. Currey (University of York, United Kingdom) has noted that mammalian antlers, auditory bullae, and a limb bone, the femur, vary in the relative amount of collagen to hydroxyapatite depending on bone function. Antlers absorb impacts, auditory bullae transmit sound, and the femur supports locomotor forces. Due to these functional differences, auditory bullae are stiff with high mineral content, while antlers have little mineralization. Limb bones are in the middle and maintain average amounts of both collagen and minerals relative to antlers and auditory bullae.

On the macroscopic scale, bone is composed of compact bone, trabecular bone, and subchondral bone. Compact bone is dense and solid, requiring a unique blood and nourishment system called the haversian system (fig. 4.1). A system of concentric bony rings (lamellae) surrounding a central canal, the haversian canal, transmits blood, nutrients, lymph, and nerves through this dense cellular region. Smaller canals, Volkmann's canals and the canaliculi, move obliquely through the lamellae to link with the haversian canal system. Living bone cells, the osteocytes, inhabit spaces within the lamellae where they make the protein collagen. Osteoblasts are bone cells that synthesize and deposit new bone tissue, while osteoclasts resorb bone tissue during remodeling. Bones are active tissues in the body that constantly respond to muscular and mechanical forces being applied

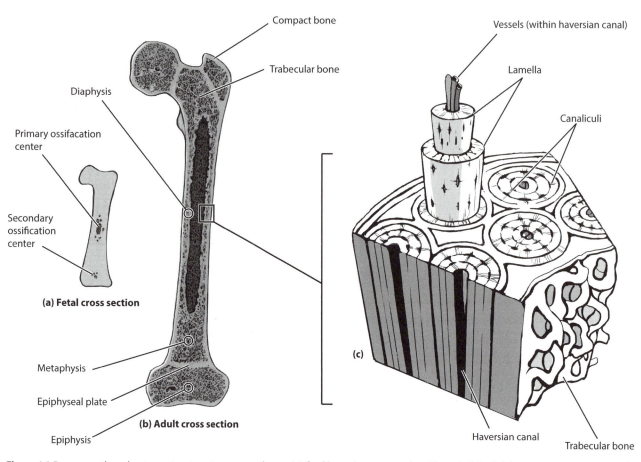

Figure 4.1 Bone growth and osteon structure in compact bone: (a) fetal bone in cross section (femur); (b) adult bone in cross section; and (c) osteon structure.

to them. Thick or dense compact bones are found where the mechanical forces are high (e.g., thick jaw bones for forceful bites during chewing cycles or in foot bones that must sustain the landing impact after a leap). In contrast, thin bones are observed in birds, which require light bodies for flight.

Trabecular bone is a porous bony structure with crisscrossing pillars that connect close to joints internally (fig. 4.1). This porous distribution of bone allows nutrient diffusion and does not require a haversian portal system for blood flow and nutrient exchange. The molecular and cellular composition of compact bone and trabecular bone is the same; only their porosity differs.

The third type of bone is called subchondral bone. Subchondral bone is found at the joint surfaces that are covered by cartilage. It is a smoother and shiner bone surface relative to non-articular surfaces and it lacks a haversian system.

Of the three embryonic germ tissues, ectoderm,

endoderm, and mesoderm, bones and muscles are derived from the mesoderm. When a mesodermal cell breaks away from the central layer of mesoderm to relocate elsewhere in the body, it eventually stops and becomes differentiated. An aggregation of mesoderm cells is called mesenchyme, the tissue that is the precursor of bone. Bone formation begins with one of two models, either endochondral or intramembranous. Endochrondral bone forms from a cartilaginous model; this pattern of bone development is generally the norm throughout the body. A cartilaginous model of a bone is constructed and then ossification gradually occurs throughout this structure, thereby forming the finished shape of a particular bone. In contrast, intramembranous bone development proceeds from a connective tissue membrane. Bone tissue gradually replaces mesenchyme but without a cartilaginous precursor. There are several primary centers of ossification within intramembranous bone, and bony spicules radiate out from these centers, making the final bone shape.

There is no difference in the kind of bone that each precursor model produces relative to these two different ossification procedures, albeit different genetic systems must be at work. Intramembranous bones are generally bones of the skull. This group includes the frontal, parietal, occipital, nasal, maxillae, vomer, and palatine bones; most of the mandible; portions of the sphenoid and temporal bones; and, surprisingly, the clavicle (part of the clavicle derives from the primitive dermal armor like the skull bones). Endochondral bones are found at the cranial base (parts of the occipital, temporal, and sphenoid bones) and the inner ear bones of the skull, and they make up almost the entire body, including the vertebrae, sternum, ribs, and limb bones. Thus, cartilage model "bones" are the precursor of most bones in the embryo.

In the embryo, blood vessels invade the cartilage precursor models and osteoblast activity begins in localized areas called ossification centers. The diaphysis, or the primary center of ossification, is located in the shaft of long bones (fig. 4.1). Bone elongation, or longitudinal growth, occurs at the epiphyseal plates located at the ends of long bones. Epiphyseal plates, or growth plates, lie between the epiphysis, a secondary ossification center, and the metaphysis, the flared end of a long bone. When the growth plates cease to grow, the epiphysis and the metaphysis fuse together and long bone lengthening ends. Skeletons are often aged by the timing of the fusion events at the epiphyseal plates.

JOINTS

Joints come in three varieties within the primate body: fibrous, cartilaginous, and synovial (fig. 4.2). Fibrous joints are not designed for movement and include joints of the skull or the interosseous membranes between the radius and ulna or the tibia and fibula. Cartilaginous joints allow a little movement but are generally connective in terms of function. These joints include the attachment region for the ribs and sternum, intervertebral disks, and the pubic symphysis. Synovial joints are the most common in the body and these joints are designed for movement. Most limb joints are synovial joints. These mobile joints are enclosed by a fibrous sheath and a joint capsule that contains synovial fluid, a lubricant secreted by

the synovial membrane lining the inside of the joint capsule (fig. 4.2). The only synovial joint in the skull is the temporomandibular joint.

Tendons, ligaments, and joint capsules help to connect joints within the musculoskeletal system. Tendons connect muscles to bones and thereby transmit the muscular contractile forces that allow joint movements to occur. Ligaments and joint capsules connect bone to bone, adding stability to joint movements and limiting the range of joint movements. Tendons and ligaments are made of connective tissue with parallel-fibered collagen, giving these structures strength and flexibility. The parallel arrangement of collagen fibers is suited to tendons, which are required to handle the unidirectional loads of muscle contractions. Ligaments, likewise, are normally parallel-oriented but they may also have collagen arrangements that are interlaced to bear loads from more than a single direction, like trabecular bone patterns. The ends of tendons and ligaments fasten onto bone like collagen fibers, first intermeshing with fibrocartilage and then gradually becoming more mineralized as these structures merge with and into a layer of cortical bone.

MUSCLES

Skeletal muscle is abundant within a mammalian body. There are more than 400 skeletal muscles arranged on both sides of a primate's body. Muscles are a soft contractile tissue made up of many individual fibers bundled together. Muscle fibers contract about one-third their total length, and a fiber's cross-sectional area determines a muscle's force capacity. Thus, long muscle fibers contract larger distances and many bundled fibers with larger cross-sectional areas produce greater forces. Parallel muscle fibers, as in the sartorius or semimembranosus muscles of the leg, represent long muscles that are able to contract and pull on their distal tendons the full one-third length of the muscle since these muscle fibers are aligned with the direction of pull. In contrast, oblique or pennate muscles, like the vastus medialis in the thigh or the gastrocnemius in the lower leg, contract at an angle to the long axis of the muscle, producing a shorter distance a tendon has to move during overall muscle contraction. Pennate muscles sacrifice distance for a greater force of contraction by adding or packing

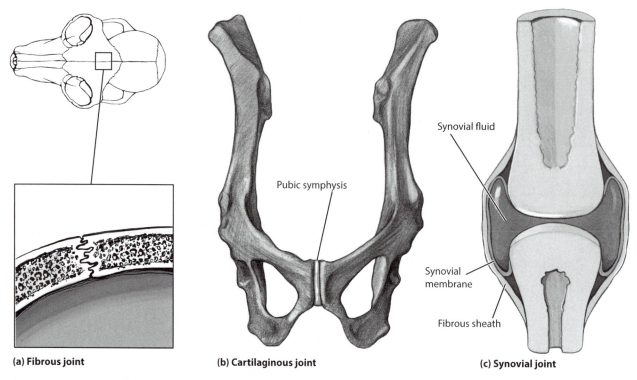

Figure 4.2 Joint types: (a) fibrous; (b) cartilaginous; and (c) synovial.

in a greater number of short muscle fibers (fig. 4.3). Muscles generally cross joints and therefore their contractions produce movements at these joints. The proximal attachment of a muscle is called its origin and its distal attachment is its insertion.

MUSCULOSKELETAL MECHANICS

Muscles frequently take part in what can be described as a musculoskeletal system of lever mechanics. Levers come in three varieties (fig. 4.4). A first-class lever is like a teeter-totter with the fulcrum in the middle. Muscular force is applied at one end to move a load at the other end. A second-class lever is like a wheel barrow or a door where the applied force is at one end, the fulcrum, is far away from the applied force, and the load is in between the two. Second-class levers are rare in a body. A third-class lever is like tweezers where the force is applied in the middle between the load and the fulcrum. Third-class levers require a greater muscular effort to move a given load but work quickly. Third-class levers are common in the mammalian body since they allow muscles to insert close to a joint they move, resulting in rapid movement with

little shortening of muscle fibers (see the discussion of velocity below). Of course, the optimum angle of pull is 90 degrees, making any other angles less effective and requiring greater muscular efforts to move a resisting load.

Muscles, when contracted, apply force and this action moves a part of a primate's body. Muscles effectively reposition bones at their respective joints. We normally think about arm and leg movements, but muscles flex, extend, and twist the trunk as well as a head or a tail. To accomplish these mechanical movements, muscles can be enhanced by lever mechanics since levers transmit force. This means that the distance (i.e., lever length) between the center of the joint (or fulcrum) and the applied muscle force is a critical factor in accomplishing the task of moving a body at its respective joints. Lever mechanics can be understood with reference to the simple equation $f_i \times l_i = f_o \times l_o$, where the "in force" ($f_i$) (the muscular force being applied) is multiplied by the "in lever" (l_i) distance (the distance to the fulcrum). This force and length value must equal the "out force" (f_o) (or load moved / the resistance) multiplied by the "out lever" (l_o) distance. In short, muscle force is applied over a

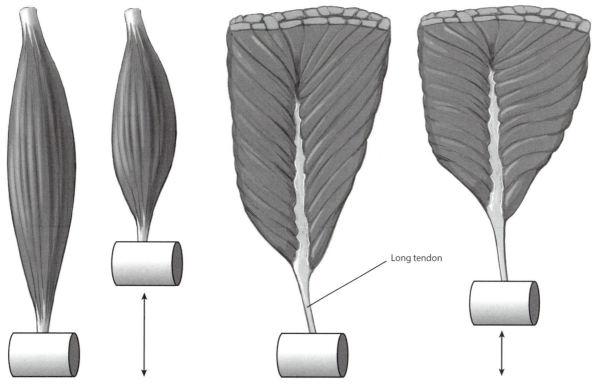

(a) Parallel muscle   **(b) Pennate muscle**

Figure 4.3 Muscle types and contraction: (a) a parallel muscle with long muscle fibers illustrating a larger distance of contraction relative to (b) a pennate muscle with compact obliquely oriented muscle fibers that provide a shorter distance of tendon contraction.

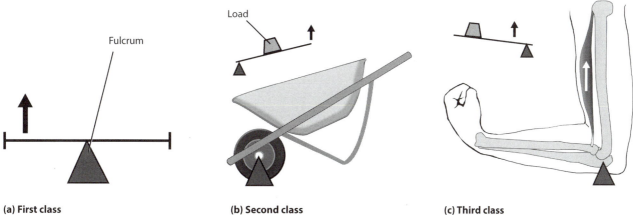

(a) First class   **(b) Second class**   **(c) Third class**

Figure 4.4 Lever types: (a) first-class lever; (b) second-class lever; and (c) third-class lever. Note the changing position of the fulcrum and lever distances in each.

certain distance on one side of a joint and this effort is carried over to the opposite side, the resistance or load side, which is the side trying to be moved or maintained by muscular effort. The equation shows that a force applied on one side of a fulcrum (or joint) must be capable of moving its opposite side if a limb is going to be moved, for example. What can a primate

do if its muscles are simply not large enough to generate enough force to move a hand-held object, for instance?

Let's consider, for example, 10 newtons (N: kg × m/s²) of muscular force being applied to move a heavier resistance load of 20 N given equal lever distances (fig. 4.5a). In this case the muscular effort

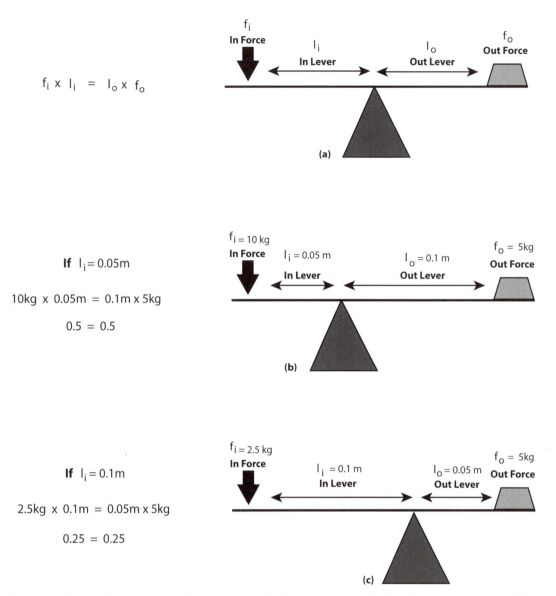

$$f_i \times l_i = l_o \times f_o$$

If $l_i = 0.05$m

10kg x 0.05m = 0.1m x 5kg

0.5 = 0.5

If $l_i = 0.1$m

2.5kg x 0.1m = 0.05m x 5kg

0.25 = 0.25

Figure 4.5 A lever example. Here the same load (5 kg) is being moved with the equation $f_i \times l_1 = l_o \times f_o$ (a, where the force and lever in values must equal the force and lever out values). (b) and (c) contrast the length of the "in lever." (c) shows that by doubling the "in lever," only one-quarter the force is needed to move the same load relative to the lever lengths in (b).

will not be sufficient to move this heavier load given the equal lever distances since the load times distance values are higher on the load side of the equation. This example shows that lever lengths need to be adjusted to improve our output so we are capable of moving a heavier load. A short "in lever" distance will require a larger force value to equal the "out force" and "out lever" side of the lever-force equation. A larger force value requires a greater muscular effort (i.e., larger muscles).

Figure 4.5b diagrams a short "in lever" distance value of 5 cm, or 0.05 m, for a first-class lever with

a longer "out lever" distance value of 10 cm (0.1 m). If the load to move is 5 kg, the "force out" and "lever out" value is equal to 0.5 N (or 5 kg × 0.1 m/s²) and this load and distance requires the "in force" value to be 10 kg of force for the 5 cm "in lever" distance to balance this equation. The fulcrum and lever position requires twice the force to move a 5-kg load within this mechanical system. What would happen if we simply changed the distance of the "in lever"? If we move the fulcrum and reverse the lever distances such that the "in lever" is now 10 cm (0.1 m), instead of 5 cm, and the "out lever" is now 5 cm (0.05 m) for the same 5-kg load, this new lever system requires a muscular effort of only 2.5 kg of force to move the same load, or one-quarter of the initial calculation value (fig. 4.5c). This simple equation and subsequent application in figure 4.5 illustrates the importance of levers and the muscular forces required to move bodies. Long "in levers" coupled with large muscles provide the best of both worlds in terms of force production.

Since muscles and joints have limited directions of motion, we can also calculate velocity of movement at joints where velocity is speed in a given direction. The velocity of bone movements is also related to the lever musculoskeletal system but in an opposite way to force. Here the lever-velocity equation $v_i \times l_o = v_o \times l_i$ or $v_o = l_o / l_i \times v_i$ has the "in velocity" being multiplied by the "out lever." The "out velocity" is overall a ratio of the two lever arms multiplied by the "in velocity." Like the lever-force equation, both sides of this equation must be equal. For high output velocities, you might notice from the lever-velocity equation that the ratio of "lever out" to "lever in" will produce a higher value when the "lever out" distance is the longer of the two distances. For high out velocities or quick joint movements, the lever-velocity equation suggests that a muscle should attach close to a joint (or fulcrum) utilizing a short "in lever" (i.e., the "in lever" should be a shorter distance relative to the "out lever" distance). For primates, a short olecranon process of the ulna means a short "in lever" for the triceps muscle, facilitating quick extension movements at the elbow (fig. 4.6). In contrast, a long olecranon process moves the insertion site away from the joint, resulting in a greater force given the increased length of the "in lever." It seems clear from the two lever equations that the same muscle and its insertion site cannot enhance both velocity and force at the same time, often described as low or high gear ratios.

Muscles that attach close to joints to move bones over a wider range of distance with a higher velocity are often described as a high gear system relative to muscles that attach away from joints (i.e., low gears; fig. 4.6). Low gears sacrifice speed but provide greater force. For primates, leapers tend to have long "out levers" and, when combined with relatively larger muscles, they are able to produce fantastic long-distance movements. When we combine long limbs

(i.e., long levers), an adaptation for long stride lengths, with large muscle power, we achieve a fast and efficient movement system. Primates with these musculoskeletal patterns often have restricted joint surfaces, for example, at the hip, a joint that constrains movements more or less in a unidirectional direction. Joints like this are unusual for primates since most primates show a pattern of highly mobile limb joints given their climbing-adapted, arboreal lifestyle.

PRIMATE LEAPING

Leaping primates need to generate high takeoff velocities, and this movement pattern represents a good case study of musculoskeletal lever mechanics. The takeoff force that is required for leaping is inversely proportional to the distance and time over which this force is applied. This means that long hindlimbs (think of long "out levers") allow primates to decrease the required force for a specific distance or increase the distance with the same musculature. Combining elongated hindlimbs with larger muscles affords primates spectacular leaping capabilities. The vastus lateralis muscle, part of the quadriceps muscle group of the thigh, is a particularly large muscle among leaping primates. Quadriceps muscles forcefully extend the lower leg at the knee as these muscles contract (fig. 4.7). This action pulls the patella upward (cranially), sliding along the patella groove of the femur, and in so doing pulls the distal patellar tendon, attached at the tibial tuberosity upward, thereby straightening or lifting the tibia relative to the femur and consequently extending the lower leg. Bent or flexed hips, knees, and ankles are all extended during a primate leap.

The knee joint of leaping primates is often quite tall (see fig. 9.14) and this increase in condylar height increases the distance from the quadriceps tendons to the center of rotation at the knee joint, providing a mechanical advantage for leg extension (fig. 4.8). Climbing primates, like lorises and apes, in contrast, have mediolaterally wide and flattened distal femora. A flatter knee reduces the distance between the patella and the knee joint, showing a decrease in the length of the "in lever" relative to the fulcrum at the knee joint. Most primate knee joints are in between these two extremes, with moderate distal femoral height.

One mechanical problem for primates of small

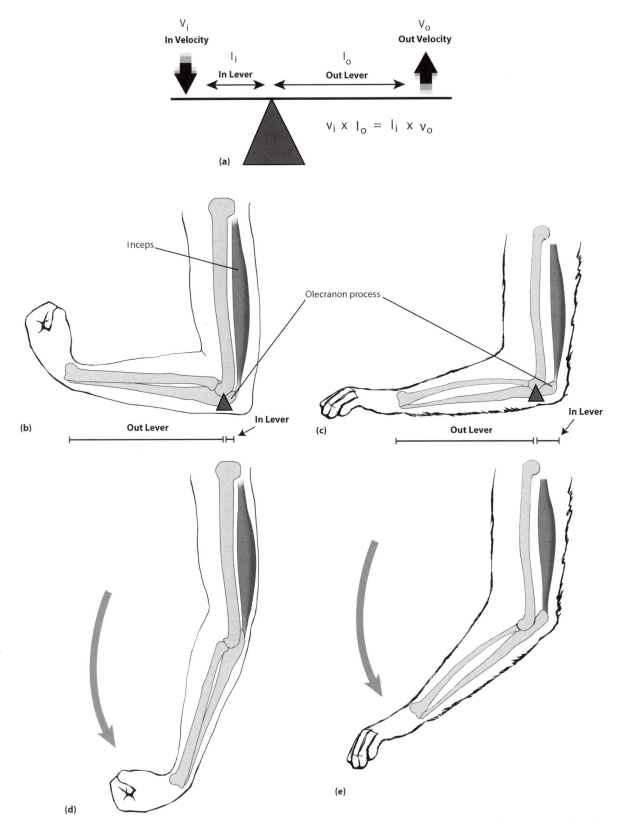

Figure 4.6 Levers and velocity. (a) illustrates the mechanical equation where "in levers" and "out levers" are on the opposite sides of this equation relative to levers and forces (see fig. 4.5). As a consequence, muscles that attach close to a joint with a short "in lever" (b) move quickly, whereas muscles that attach away from joints (c) move slower but with more force. In (d) and (e), the attachment points (insertions) of the triceps muscle is closer (b) or farther (c) from the joint and this implies greater extension for (d) relative to (e).

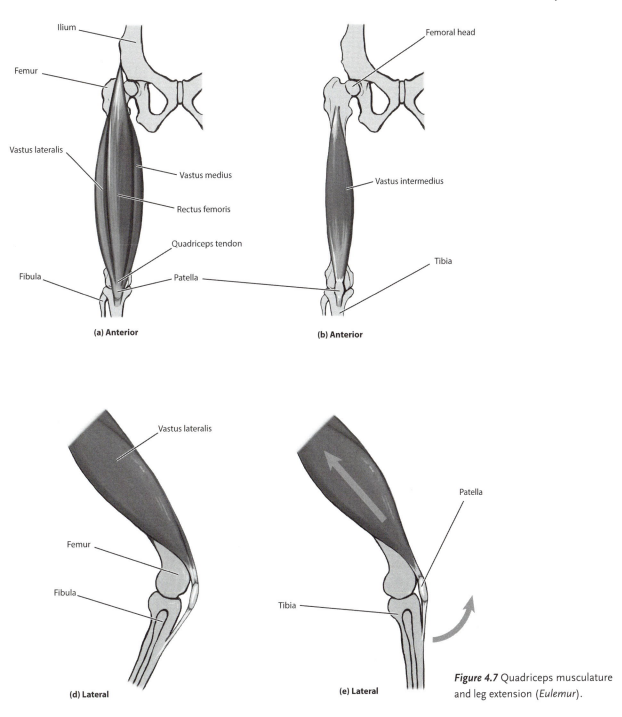

Figure 4.7 Quadriceps musculature and leg extension (*Eulemur*).

size is generating enough propulsive force given their much smaller-sized musculature. For example, the vastus lateralis muscle in a human thigh is a far larger muscle than this same muscle in a tarsier, yet tarsiers are great leapers while humans are not. A large muscle can certainly produce a larger force relative to a smaller muscle on the basis of its cross-sectional area,

so why are tarsiers better leapers than we? In fact, tarsiers are capable of leaping several times (usually five or more times) their entire body length relative to a human leap (e.g., a standing broad jump for humans is normally one body length). Small primates, like tarsiers, are simply better leapers due to their small size, their elongated lower limbs, and their relatively

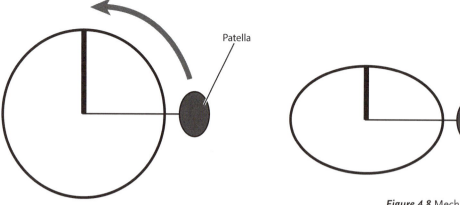

(a) Tall Knee   **(b) Flat Knee**

Figure 4.8 Mechanical advantage (lever distance difference) of tall versus flat knees in primates.

large quadriceps muscles. Both adaptations, leg length and muscle size, provide better lever and propulsive mechanics to propel tarsiers forward in comparison to humans, where larger body sizes (volume), relatively smaller muscles in proportion to body size (muscle cross-sectional areas to volume ratios; see below), and differing lever mechanics in human legs (e.g., short "out levers" in the foot) constrain overall leaping distances.

SIZE

The most important constraint in the tarsier to human comparison is body size, a volume or cubed variable. Since weight increases at a greater pace relative to the cross-sectional area of the propulsive muscles (a squared variable), large-sized animals must generate even higher propulsive forces from their musculature to offset the area-to-volume ratio imbalance (fig. 4.9). This ratio predicts that large-sized primates will reach a point where leaping is too mechanically difficult and inefficient to move a body forward; plus, large size adds considerable risk. Anthropoids fall into this size dilemma more often than do the small tarsiers and strepsirhine primates. In fact, the ceiling for frequent leaping anthropoids is around 10 kg. A large galago can leap as far as a small galago, but to achieve this distance the large galago needs to move more mass to cover the same distance, requiring greater leg musculature force and a better takeoff angle. The optimal takeoff angle is 45 degrees but most primates utilize lower takeoff angles with flatter trajectories. General-

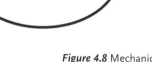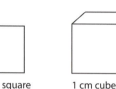

1 cm square
Area = 1 x 1 = 1   1 cm cube
Volume = 1 x 1 x 1 = 1   2 cm cube
Volume = 2 x 2 x 2 = 8

Side Length	Area	Volume	Area/Volume
1cm	1cm²	1 cm³	1
2	4	8	1/2
3	9	27	1/3
4	16	64	1/4

Figure 4.9 Area-to-volume relationship. The area-to-volume ratio decreases with increasing volume (size),

ized leapers can leap as far as specialized leapers but it takes more effort.

Consider the locomotor problem that the largest arboreal mammal, the orangutan, a species where large males weigh around 90 kg, has while moving in the upper canopy. Orangutans move cautiously and slowly in the upper canopy with lots of suspensory, climbing, and bridging movements. Orangutans are well adapted for life in the trees, but they try to spread their immense body weight over multiple arboreal supports and they test supports before transferring their full body weight onto them. If orangutans were leaping primates, only a few branches might be able to withstand the impact of their large bodies upon landing. If orangutans fall, their large size (or volume, a cubed variable) relative to their bone strength (cross-sectional area, a squared variable) implies a high risk for bone breakage and injury. As size increases more steeply relative to area, area-to-volume ratios decrease for large animals, leaving them at greater risk for injury (fig. 4.9). Professor J.B.S. Haldane notes: "You can drop a mouse down a thousand-yard mine shaft; on arriving at the bottom, it gets a slight shock and walks away, provided that the ground is fairly soft. A rat is killed, a man is broken, a horse splashes."* In point of fact, large male orangutans make long-distance traveling events along the ground, decreasing the likelihood of injury from a fall from the upper canopy.

BRACHIATION

Besides leaping, there are several other important musculoskeletal adaptations across primates to consider. There are two highly derived adaptations among primates. The first is the movement called brachiation, an unusual arm-swinging motion found in the living apes and humans; it involves important adaptations of the shoulder that contrast with the shoulder form of other primates (see chapter 8). The second, the grasping foot, is one of the most fundamental of all primate musculoskeletal adaptations.

The shoulder joint of living apes is composed of three bones (clavicle, scapula, and humerus) and 17 muscles, making this joint a complicated musculoskeletal adaptive system. The anatomical position of ape and human shoulders is unusual (see figs. 7.10 and 8.9). In these primates, the shoulders are pushed onto the far sides, toward the back, of the thorax due to the dorsoventrally compressed and laterally wider chests of apes and humans (see chapter 7). This back scapular position necessitates a long clavicle. The ape clavicle articulates with an enlarged acromial process of the scapular spine while a large, ball-like joint surface with a medially rotated humeral head articulates with the glenoid fossa of the scapula. Note that a primate humerus simply hangs, via ligaments and muscles, from the scapula, forming a ball and socket joint like that of the hip but without the encapsulating bony socket of the hip joint. This anatomical arrangement is good for mobility but is certainly less protected and less stable

---

* Haldane, J.B.S. 1927. On being the right size; p. 1 *in* Possible Worlds and Other Essays. Chatto and Windus, London.

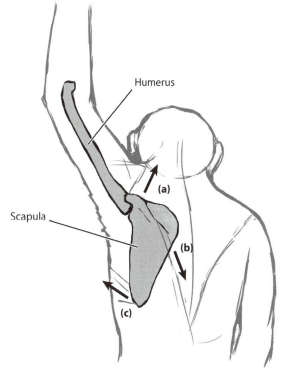

Pan (posterior)

Figure 4.10 Scapular mobility in ape (*Pan*) shoulders. The points a, b and c show the upward rotation of the scapula via the trapezius (a, b) and the serratus anterior (c) muscles.

compared to the hip joint. The laterally positioned shoulder of apes and humans allows the humerus to be raised and rotated in a circular motion above the head, a motion called circumduction (fig. 4.10), and an arm position adapted for suspension and brachiation (arm swinging) in apes.

Several muscles are involved in producing this complicated shoulder motion (fig. 4.11). You may have heard of one key muscle group if you follow major league baseball. This muscle group is called the rotator cuff, which consists of four muscles: the subscapularis, the supraspinatus, the infraspinatus, and the teres minor. These four muscles rotate the humerus. The infraspinatus and teres minor muscles rotate the humerus laterally, while the subscapularis (with the teres major) rotates the humerus medially. The supraspinatus, along with the deltoid muscle, begins the motion of circumduction by abducting and lifting the humerus above the head. To achieve a maximum elevated arm position, the scapula also needs to rotate upward, or cranially, via the trapezius and serratus

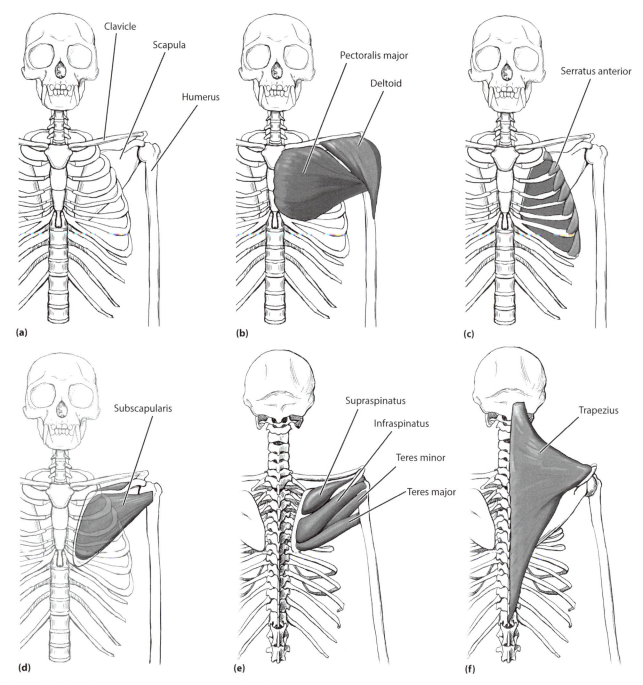

Figure 4.11 Shoulder anatomy and muscles in apes (*Hylobates*).

anterior muscles of the thorax. The pectoralis major helps to flex the humerus upon arm lifting.

It takes a rather complicated series of muscle contractions to achieve a shoulder movement humans and living apes are able to do from a young age. In contrast, lemurs, monkeys, and other mammals are unable to produce this over-the-head rotational motion (see chapter 8). In terms of muscle function,

the shoulder and upper thorax muscles involved in circumduction tend to be relatively short, with oblique fiber patterns and without long tendons. Most insert close to the joint, producing fast gear movements, as noted above. Four muscles—the teres major, the serratus anterior, the deltoid, and the pectoralis major—insert farther away from the shoulder joint, or fulcrum, and they in contrast reflect a low gear system

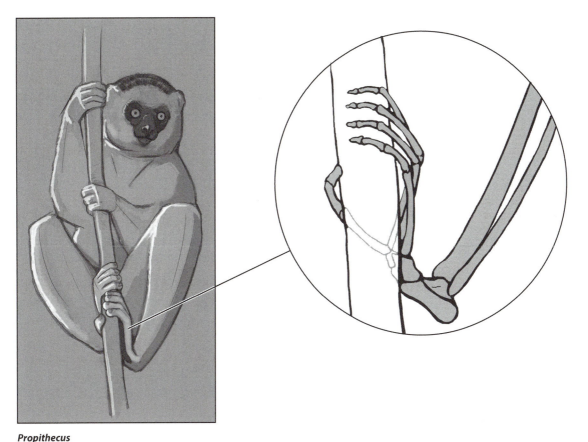

Propithecus

Figure 4.12 Foot grasping in primates.

designed to produce greater muscular force (fig. 4.10). Overall, the shoulder joint and its corresponding musculature within living apes and humans is quite a novel adaptation relative to that of the more quadrupedally oriented primates.

GRASPING FOOT

The primate grasping foot is the most fundamental adaptation of the entire Order Primates (fig. 4.12). In the primate foot all five digits and one cuneiform bone, the entocuneiform, are intimately involved in digital grasping (fig. 4.13). The four lateral digits simply flex at the interphalangeal and metatarsophalangeal joints via the flexor digitorum longus, the flexor digitorum brevis, and a few intrinsic muscles. All of the curved digital phalanges are able to flex, or bend, around and hold on to a curved substrate. The movement of the big toe is more complex in that its normally abducted position allows the first digit to swing across the sole to be in opposition to the lateral digits during grasp-

ing, with digital pressure being applied at the nailed tips of the toes, the distal phalanges. Primate grasping is especially designed for small-diameter curves (i.e., small arboreal branches).

The saddle-shaped joint of the distal entocuneiform is the key joint for big toe, or hallucal, grasping. This joint articulates both the entocuneiform and the proximal joint of the first metatarsal; together these two joint surfaces allow a mediolateral, or swing, motion to occur across the sole by the big toe (fig. 4.13). The swing movement is accomplished by the contraction of the adductor hallucis and peroneus longus muscles (i.e., adduction of the first digit), while the long and short flexor tendons (the flexor hallucis longus and the flexor hallucis brevis) produce digital flexion at the first digit. The flexor hallucis longus and the adductor hallucis can be particularly large muscles, thereby generating forceful first digit grasps in primates (fig. 4.14).

In addition to grasping, the primate foot must be able to rotate inward (foot inversion) toward the

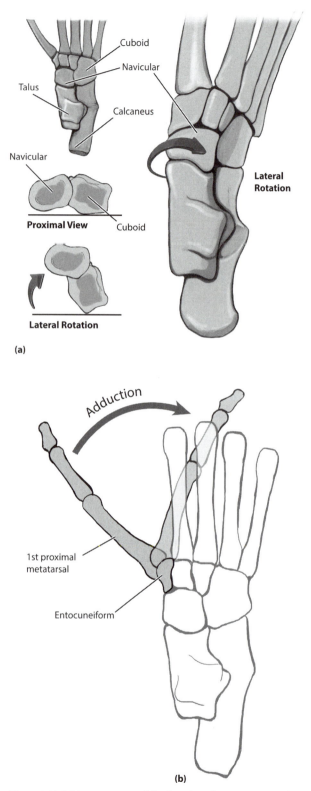

Figure 4.13 Adductive swing of the first digit during grasping (b) and lateral rotation at the transverse tarsal joint to accomplish foot inversion in primates. The navicular and cuboid tarsal bones must both rotate laterally to complete this foot position.

curved surface of a branch while grasping. This is a novel adaptation among mammals requiring great foot mobility and many muscles, making primate feet heavy and muscular relative to the feet of running mammals. To accomplish this rotation, primate feet rotate between their tarsal bones, especially along the transverse tarsal joint and the subtalar joint (fig. 4.13). In contrast to the primate foot, rotation of the primate hand does not occur between the homologous carpal bones in the wrist; instead, this motion occurs at the elbow between the capitulum of the humerus and the radial head of the radius.

The four bones of the primate transverse tarsal joint are the talus, the navicular, the calcaneus, and the cuboid. The talar head and the curved proximal navicular joint form one joint surface, the talonavicular joint, and together they represent half of the transverse tarsal joint; the other half occurs at the calcaneocuboid joint. To invert a primate foot, both the navicular and cuboid bones must rotate laterally (fig. 4.13). To begin this rotation, the invertor musculature, primarily tibialis anterior and both of the long flexors (the flexor digitorum longus and the flexor hallucis longus), contract and pull on the medial part of the plantar side of the foot. The contraction of the tibialis anterior pulls the navicular laterally so that it lies on top of the cuboid, rather than being side to side. This contraction begins foot inversion. Next the long flexor tendons contract and they pull the distal parts of the foot toward the talus and calcaneus at the transverse tarsal joint. These proximal movements force the cuboid (i.e., the cuboid process) to screw into the depressed calcaneal pit surface in the calcaneocuboid joint surface. As the cuboid screws inward, this motion allows the cuboid to slide closer to the talonavicular joint and in effect allows the calcaneocuboid joint to be in a transverse line with the talonavicular joint (i.e., the transverse tarsal joint). This four-tarsal arrangement facilitates the lateral rotation of the foot during foot inversion.

Additionally, the long flexor tendons help to rotate the calcaneus laterally as they contract since their tendons pull up from below the sustentaculum tali along the medial side of the calcaneus while the short calcaneocuboid ligaments keep both the calcaneus and cuboid bones coupled together. As the calcaneus

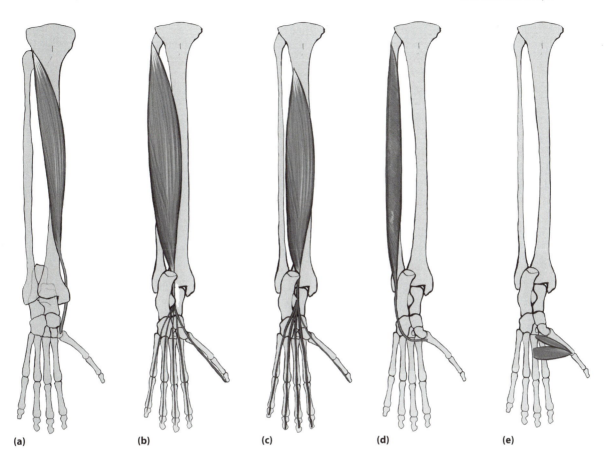

(a)   **(b)**   **(c)**   **(d)**   **(e)**

Figure 4.14 Important grasping muscles in the leg and foot of primates (*Eulemur*): (a) tibialis anterior; (b) flexor hallucis longus; (c) flexor digitorum longus; (d) peroneus longus; (e) adductor hallucis.

rotates laterally, the talus is shifted back along the subtalar joint and the talar head is lifted above and medially away from the sustentaculum tali. All of this reverses to evert the primate foot. This complicated series of muscle contractions and bone movements allow primate feet to grab on to a branch and move up or along a curved surface with speed and agility within the complex three-dimensional array of canopy substrates that vary in size and orientation. Grasping adaptations in the foot and hand have allowed primates to master arboreal niches in high- and low-canopy environments.

Selected References

Aiello, L., and C. Dean. 1990. An Introduction to Human Evolutionary Anatomy. Academic Press, New York.

Alexander, R.M. 1981. Animal Mechanics. Blackwell Scientific Publications, Oxford, UK.

———. 2003. Principles of Animal Locomotion. Princeton University Press, Princeton, N.J.

Anemone, R.I. 1990. The VCL hypothesis revisited: patterns of femoral morphology among quadrupedal and saltatorial prosimian primates. American Journal of Physical Anthropology 83:373–393.

Ashton, E.H., and C.E. Oxnard. 1963. The musculature of the primate shoulder. Transactions of the Zoological Society of London 29:553–650.

Basmajian, J.V. 1976. Primary Anatomy. 7th edition. Williams and Wilkins Company, Baltimore.

Berringer, O.R., F.M. Browning, and C.R. Schroeder. 1968. An Atlas and Dissection Manual of Rhesus Monkey Anatomy. Anatomy Laboratory Aids, Florida State University, Tallahassee.

Carter, D.R., and G.S. Beaupré. 2001. Skeletal Function and Form—Mechanobiology of Skeletal Development, Aging and Regeneration. Cambridge University Press, Cambridge, UK.

Crompton, R.H., and W.I. Sellers. 2007. A consideration of leaping locomotion as a means of predator avoidance in prosimian primates: pp. 127–145 *in* S. Gursky and K.A.I.

Nekaris (eds.), Primate Anti-Predator Strategies. Springer, New York.

Crompton, R.H., W.I. Sellers, and M.M. Gunther. 1993. Energetic efficiency and ecology as selective factors in the salutatory adaptation of prosimian primates. Proceedings of the Royal Society of London [Biology] 254:41–45.

Currey, J.D. 1984. The Mechanical Adaptations of Bones. Princeton University Press, Princeton, N.J.

———. 2006. Bone: Structure and Mechanics. Princeton University Press, Princeton, N.J.

Demes, B., J.G. Fleagle, and P. Lemelin. 1998. Myological correlates of prosimian leaping. Journal of Human Evolution 34:385–399.

Demes, B., W.L. Jungers, J.G. Fleagle, R.F. Wunderlich, B.G. Richmond, and P. Lemelin. 1996. Body size and leaping kinematics in Malagasy vertical clingers and leapers. Journal of Human Evolution 31:367–388.

Diogo, R., J.M. Potau, J.F. Pastor, F.J. de Paz, E.M. Ferrero, G. Bello, M. Barbosa, M.A. Aziz, A.M. Burrows, J. Arias-Martorell, and B.A. Wood. 2012. Photographic and Descriptive Musculoskeletal Atlas of Gibbons and Siamangs (Hylobates). CRC Press, Boca Raton, Fla.

Diogo, R., J.M. Potau, J.F. Pastor, F.J. de Paz, E.M. Ferrero, G. Bello, M. Barbosa, and B.A. Wood. 2010. Photographic and Descriptive Musculoskeletal Atlas of Gorilla. Taylor and Francis, Oxford, UK.

Diogo, R., and B.A. Wood. 2012. Comparative Anatomy and Phylogeny of Primate Muscles and Human Evolution. CRC Press, Boca Raton, Fla.

Erikson, G.E. 1963. Brachiation in the New World monkeys; pp. 135–165 in J. Napier and N.A. Barnicot (eds.), The Primates. Symposia Zoological Society of London, No. 10.

Fleagle, J.G. 1974. Dynamics of brachiating siamang (Hylobates symphalangus syndactylus). Nature 248:259–260.

Gebo, D.L. 1987. Functional anatomy of the tarsier foot. American Journal of Physical Anthropology 73:9–31.

Gebo, D.L. 1993. Functional morphology of the foot in primates; pp. 175–196 in D.L. Gebo (ed.), Postcranial Adaptation in Nonhuman Primates. Northern Illinois University Press, DeKalb.

Gebo, D.L. 2011. Vertical clinging and leaping revisited: vertical support use as the ancestral condition of strepsirhine primates. American Journal of Physical Anthropology 146(3):323–335.

Hall-Craggs, E.C.B. 1965. An analysis of the jump of the lesser galago. Journal of Zoology (London) 147:20–29.

———. 1966. Rotational movements in the foot of Galago senegalensis. Anatomical Record 154:287–294.

Hartman, C.G., and W.L. Straus (eds.). 1933. The Anatomy of the Rhesus Monkey. Hafner Publishing Company, New York.

Hildebrand, M. 1974. Analysis of Vertebrate Structure. John Wiley and Sons, New York.

Jouffroy, F.K. 1962. La musculature des members chez les Lémuriens de Madagascar: etude descriptive et comparative. Mammalia 26:1–326 (Supplement 2).

Jungers, W.L. (ed.). 1985. Size and Scaling in Primate Biology. Plenum Press, New York.

Jungers, W.L., and J.T. Stern. 1980. Telemetered electromyography of forelimb muscle chains in gibbons (Hylobates lar). Science 208:617–619.

Larson, S.G. 1988. Subscapularis function in gibbons and chimpanzees: implications for interpretation of humeral head torsion in hominoids. American Journal of Physical Anthropology 76:449–462.

———. 1993. Functional morphology of the shoulder in primates; pp. 46–69 in D.L. Gebo (ed.), Postcranial Adaptation in Nonhuman Primates. Northern Illinois University Press, DeKalb.

Larson, S.G., J.T. Stern, and W.L. Junger. 1991. EMG of serratus anterior and trapezius in the chimpanzee: scapular rotators revisited. American Journal of Physical Anthropology 85:71–84.

Lewis, O.J. 1980a. The joints of the evolving foot. Part 1. The ankle joint. Journal of Anatomy 130:527–543.

———. 1980b. The joints of the evolving foot. Part 2. The intrinsic joints. Journal of Anatomy 130:833–857.

———. 1981. Functional morphology of the joints of the evolving foot. Symposia of the Zoological Society of London 46:169–188.

———. 1989. Functional Morphology of the Evolving Hand and Foot. Clarendon Press, Oxford, UK.

Martin, R.D. 1990. Primate Origins and Evolution—A Phylogenetic Reconstruction. Princeton University Press, Princeton, N.J.

Morton, D.J. 1922. Evolution of the human foot, part 1. American Journal of Physical Anthropology 5:305–336.

———. 1924. Evolution of the human foot, part 2. American Journal of Physical Anthropology 7:1–52.

Nordin, M., and V.H. Frankel. 1989. Basic Biomechanics of the Musculoskeletal System, 2nd edition. Lea and Febiger, Philadelphia.

Raven, H.C. (arranged and edited by W.K. Gregory). 1950. The Anatomy of the Gorilla. Columbia University Press, New York.

Schön, M.A. 1968. The Muscular System of the Red Howling Monkey. Bulletin 273:1–85. Smithsonian Institution Press, Washington, D.C.

Stern, J.T. 1971. Functional Myology of the Hip and Thigh of Cebid Monkeys and its Implications for the Evolution of Erect Posture. Bibliotheca Primatologica, No. 14. S. Karger, Basel, Switzerland.

Swartz, S. 1993. Biomechanics of Primate Limbs; pp. 5–42 in D.L. Gebo (ed.), Postcranial Adaptation in Nonhuman Primates. Northern Illinois University Press, DeKalb.

Swindler, D.R., and C.D. Wood. 1982. An Atlas of Primate Gross Anatomy (Baboon, Chimpanzee, and Man). R.E. Krieger Publishing Company, Malabar, Fla.

Szalay, F.S., and M. Dagosto. 1988. Evolution of the hallucal grasping in the primates. Journal of Human Evolution 17:1–33.

Szalay, F.S., and R.L. Decker. 1974. Origins, evolution and function of the tarsus in Late Cretaceous eutherians and Paleocene primates; pp. 223–259 *in* F.A. Jenkins (ed.), Primate Locomotion. Academic Press, New York.

Tuttle, R.H. 1972. Relative mass of cheiridial muscles in catarrhine primates; pp. 262–291 *in* R.H. Tuttle (ed.), Function and Evolutionary Biology of the Primates. Aldine-Atherton, Chicago.

Tuttle, R.H., and J.V. Basmajian. 1976. Electromyography of pongid shoulder muscles and hominoid evolution. I. Retractors of the humerus and rotators of the scapula. Yearbook of Physical Anthropology 20:491–497.

Tuttle, R.H., and J.V. Basmajian. 1978. Electromyography of pongid shoulder muscles. II. Deltoid, rhomboid, and "rotator cuff." American Journal of Physical Anthropology 49:47–56.

5 | Heads

BONES OF THE SKULL

Mammalian skulls are symmetrical by design and by definition include the cranium and the mandible. The bony architecture of mammalian skulls is complicated due to the number of bones that comprise a skull and the hidden nature of several bones enclosed within the cranium. Skull morphology is uniquely associated with each Order of Mammalia. The bones that make up a primate skull are homologous with those of other mammals, including humans, meaning that once you learn these skeletal elements they are readily identifiable across Mammalia. The primate skull can be broken down into five osteological regions: the face and nasal region (rostrum); the braincase; the basicranium; the internal mid-skull region; and the mandible.

The face and nasal region contains the nose and orbits of primates as well as holding all of the upper dentition (fig. 5.1). Several smaller bones of the skull are also located in this region. Beginning with the distal end of the rostrum, we encounter the premaxilla and the maxilla, two bones with right and left halves that contain the upper dentition. Both of these bones form the bony edge that surrounds the bottom and sides of the nasal aperture. At the top of the nasal aperture, the maxilla meets the right and left nasals that articulate with the frontal bone of the skull. Rostral bones, including the nasal bones, are longer among the long-nosed primates (e.g., strepsirhines and baboons). Within the nasal aperture lie several small turbinate bones (fig. 5.2) involved with the sen

sory epithelium of smelling (i.e., the maxilloturbinate, nasoturbinate, and several ethmoturbinates). Turbinate bones are more numerous and complex in shape among the wet-nosed, or more smelling-oriented, primates. Two bony perpendicular plates, one each from the ethmoid and vomer, separate the nasal aperture into right and left halves. The ethmoid bone is best viewed from a skull with its top removed, while the vomer is more easily identified from an upside down view at the back of the palate. The ethmoid is located within the ethmoid notch between the orbits of the frontal bone (fig. 5.3). The ethmoid is a central element in the transfer of nerves from the nasal region to the olfactory bulb (cranial nerve I) of the brain. These many separate nerves carry the sensory information for smell and enter the ethmoid through a structure called the cribiform plate, a bony structure that resembles a sieve (many foramena pierce this bone). Primates that depend on a keen sense of smell (i.e., the strepsirhine primates) have elaborated this bone.

Going back to the major bones of the face (fig. 5.1), the right and left maxillae surround the nasal aperture; they also form the bottom of the two orbits, and continue along the side of the primate skull where the maxillae articulate with the zygomatics, or cheek bones, laterally. This is an important anatomical region as primates have evolutionarily shifted their orbits to face in a more forward direction relative to other mammals, which retain laterally oriented eye positions. The zygomatic and frontal bones of primates form a postorbital bar, a ring of bone that

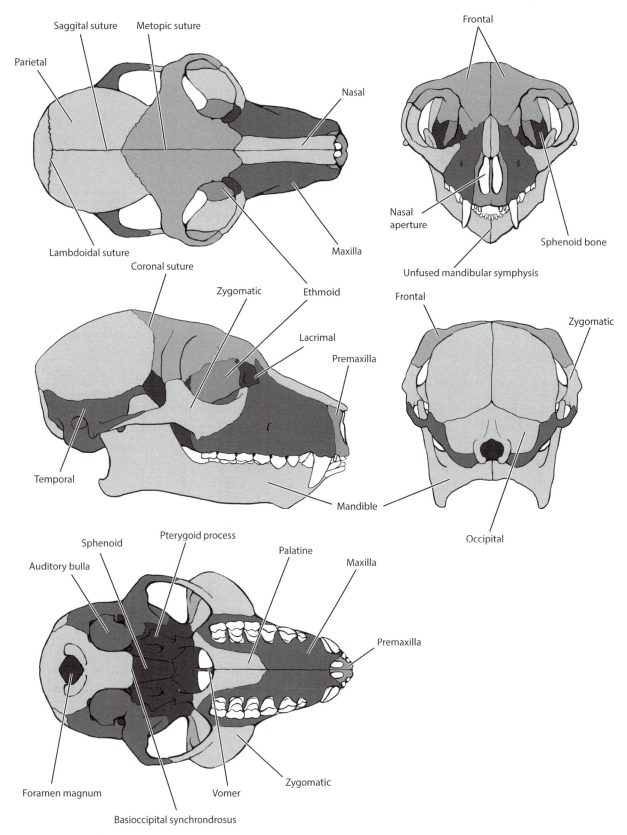

Figure 5.1 Bones and features of the primate skull (*Eulemur fulvus*).

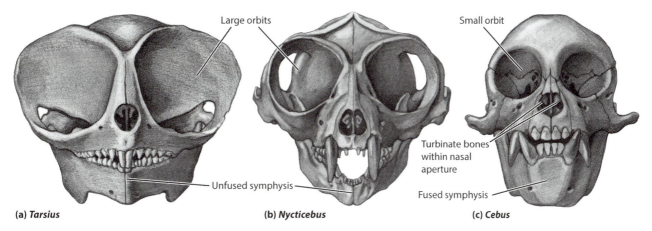

Figure 5.2 Frontal view of three primate skulls: (a) *Tarsius*; (b) *Nycticebus*; and (c) *Cebus*. Note the enlarged orbits on the taxa to the left of *Cebus*.

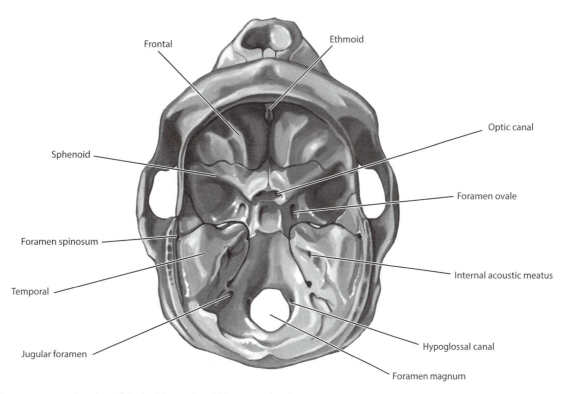

Figure 5.3 A cut-away superior view of the inside cranium (*Macaca mulatta*).

circles around the eye orbit, found among strepsirhine primates (fig. 5.4). In tarsiers, the frontal bone and a part of the sphenoid, the alisphenoid, have expanded the postorbital bar with extra bony flanges relative to the structure found in strepsirhines. In contrast, anthropoids have completely walled off the back of the orbit by fusing parts of the frontal, zygomatic, and sphenoid bones. We call this anthropoid condition postorbital closure and it represents a key distinc-

tion between anthropoid skulls and those of other primates.

Six bones—the frontal, ethmoid, lacrimal, maxilla, zygomatic, and sphenoid bones—meet to form the outer casing (or orbit) that holds the primate eye. There are fissures and foramina (e.g., the optic canal) that allow soft tissue entry into the orbit.

Behind the orbits lies the braincase (see fig. 5.1). The primate braincase includes the large, curved

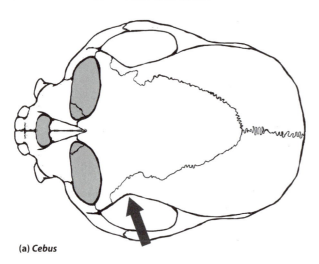

(a) *Cebus*

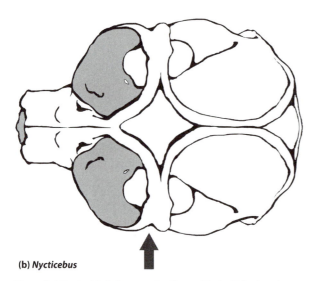

(b) *Nycticebus*

Figure 5.4 Postorbital closure of anthropoids (a, *Cebus*) and a postorbital bar in strepsirhines (b, *Nycticebus*).

the temporalis muscle, attaches to the sides of each parietal (fig. 5.5). All three of these large cranial bones enclose and protect the primate brain.

Along the sides of a primate skull (see fig. 5.1) are two cranial bones, the temporal bone and the sphenoid, often divided into two bones among primates. The temporal bone extends a bony process forward to connect with the zygomatic, thereby forming the zygomatic arch. Here the masseter muscle, another important chewing muscle, originates.

The sphenoid is a complicated bone that pokes out and is visible within the orbit, along the basicranium and at the sides of the primate skull, but its main location is within the skull behind the orbits, where it is connected to several different cranial bones (figs. 5.1 and 5.3), 12 in total. The sphenoid is where the optic

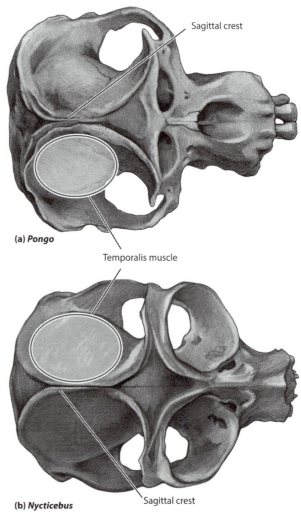

Sagittal crest

(a) *Pongo*

Temporalis muscle

(b) *Nycticebus*

Sagittal crest

Figure 5.5 Sagittal crest formation for enlarged temporalis muscles in primates.

bones at the top and sides of the skull, beginning in the front with the frontal bone—actually two separate bones in strepsirhines and tarsiers, but only a single bony element among anthropoids. The frontal bone, besides being at the top front of the skull, also represents the top part of each orbit. The frontal articulates along its sides with the sphenoid and from behind with the right and left parietals. The parietals are followed in turn by the occipital bone at the rear of the skull. Besides being the back half of a primate skull, the occipital bone is where the jugular vein exits the skull to transfer venous blood back to the heart, where the spinal cord enters the skull, and where the muscles of the neck attach along the nuchal plane of the occipital bone. In a similar way, an important chewing muscle, nerves cross, the hypothalamus sits, and the pterygoid chewing muscles attach. To appreciate the sphenoid, you need to look inside a skull and examine the endocranium.

Behind the sphenoid on the lateral side of the cranium lies the temporal bone (see fig. 5.1), another complicated skull element like the sphenoid. The temporal bone contains the temporomandibular joint, where the cranium and the mandible articulate, as well as the bony elements for the ear, including the external auditory meatus. The temporal bone also contains the auditory bulla with its three tiny inner ear ossicles: the stapes, the malleolus, and the incus. Surrounding the auditory bulla is where the internal carotid artery enters a foramen to deliver blood to the brain.

If we turn a primate skull upside down, this view provides a direct look at the upper dentition (see chapter 6) and the hard palate, a bony structure made up of the premaxilla, the maxilla, and the palatine bones, lefts and rights for all (fig. 5.6; see also fig. 5.1). This view also shows four pterygoid processes, or plates, projecting away from the sphenoid bone, being

the site of origin for the pterygoid chewing muscles. This basicranial view also allows us to examine how the sphenoid bone, a key element in the mid-skull internal region, articulates with the occipital bone at the basioccipital synchrondrosus, and where the spinal cord enters the skull through the foramen magnum, a large hole in the occipital bone (figs. 5.1 and 5.6).

The primate mandible, or jaw, is composed of a single bone but with right and left halves among strepsirhines and tarsiers, being unfused and attached together anteriorly by ligaments at the symphysis (see fig. 5.2). In anthropoids, the symphyseal region is fused and only a single bony element is present. The primate mandible holds all of the lower dentition (see chapter 6).

STREPSIRHINE HEADS

If you place a series of primate and mammal skulls on a table in a classroom setting and ask students to separate these skulls, having known nothing about these mammals initially, they soon realize that some skulls have large braincases. These are the primates, and all

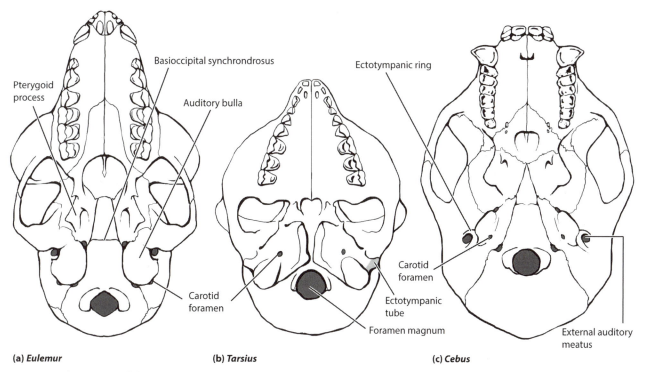

Figure 5.6 Inferior views of the primate skull base. Note the different anatomical locations of entry for the internal carotid artery through the carotid foramen across primates.

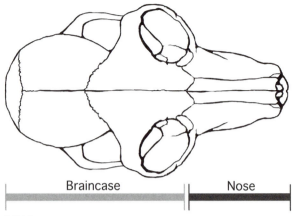

Braincase | Nose

(a) *Eulemur*

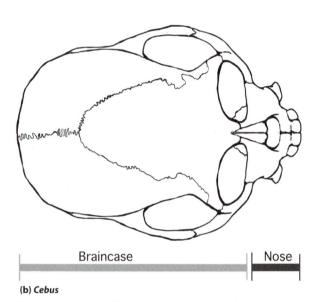

Braincase | Nose

(b) *Cebus*

Figure 5.7 Primate skull types: (a) a strepsirhine skull, *Eulemur*, with a long nose and a relatively shorter braincase; and (b) an anthropoid skull, *Cebus*, with a short nose and a large braincase.

have enlarged brains relative to those of other mammals, especially at comparable body sizes. When we begin to sort the primate skulls, we see two distinctly different head shapes across the living primates that involve noses and brains. The primitive head type for primates has a long nose and a relatively smaller brain and is found among the strepsirhine primates. The other head type has a much larger brain but with a short nose; this version is found among the haplorhine radiation of primates (fig. 5.7).

In strepsirhine primates (figs. 5.8 and 5.9), the nose is long and well developed internally for olfaction (smelling), the braincase is smaller, and the eyes are oriented obliquely for orbital convergence with a ring

of bone surrounding each orbit (i.e., a postorbital bar; see fig. 5.4). The long noses of strepsirhines are filled with small turbinate bones that are lined with special sensory nerves within the internal nasal aperture (see fig. 5.2). These nasal adaptations imply a strong sense of smell and smelling-oriented behaviors (i.e., scent marking) in strepsirhines relative to the short-nosed haplorhines. Nose length varies by strepsirhine group, with lemurs possessing the longest noses and aye-ayes the shortest.

Nocturnal strepsirhines are renowned for their enlarged orbits and large eyes relative to daytime, or diurnal, primates (see fig. 5.2). Large eyes help increase light absorption in the limited light environment at night.

Like smelling, hearing is a key ability for primates adapted for night living. Strepsirhine ears can be tall and pointed or small and round in terms of soft tissue shape but all have a ring of bone, the ectotympanic bone, just inside the ear opening (i.e., the external auditory meatus). The eardrum is stretched across this ring attaching within the auditory bulla, a structure made from a single bone, the petrosal, across all primates. Primatologists often use the position of the ectotympanic bone to identify taxonomic groups (see fig. 5.6). For example, in lemurs the ectotympanic ring is suspended within the auditory bulla whereas the ring is attached outside of the bulla among the lorises and galagos.

There are several other specific skull features that strepsirhine skulls share in common. One is the foramen for the internal carotid artery, a major arterial supplier to the brain. This artery enters the auditory bulla posterolaterally in strepsirhines (see fig. 5.6). The internal carotid supplies blood to the brain and to the eyes, along with the vertebral arteries. The internal carotid artery splits into two branches (fig. 5.10), the stapedial artery, which runs through the stapes, one of the three inner ear bones, and the promontory artery; both arteries enter the brain. The promontory branch fuses with the circle of Willis; this artery is the major arterial blood source for the brain. Lorises, galagos, and cheirogaleids use an alternative carotid source to supply the brain in contrast to lemurs. Here the common carotid splits and a branch of the external carotid artery, the ascending pharyngeal artery, is the major arterial supplier for the brain in these groups.

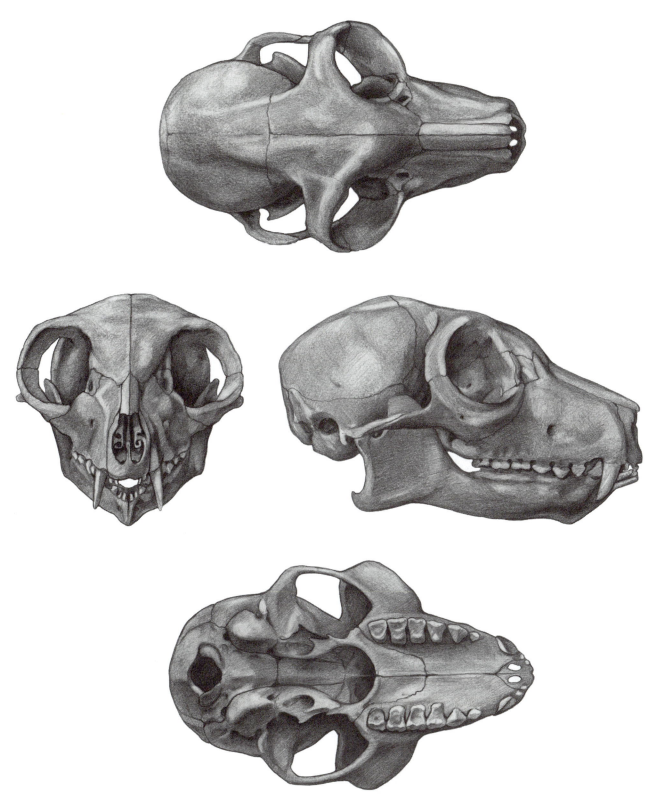

Figure 5.8 Four views of the skull of *Eulemur fulvus* (top, front, side, and below).

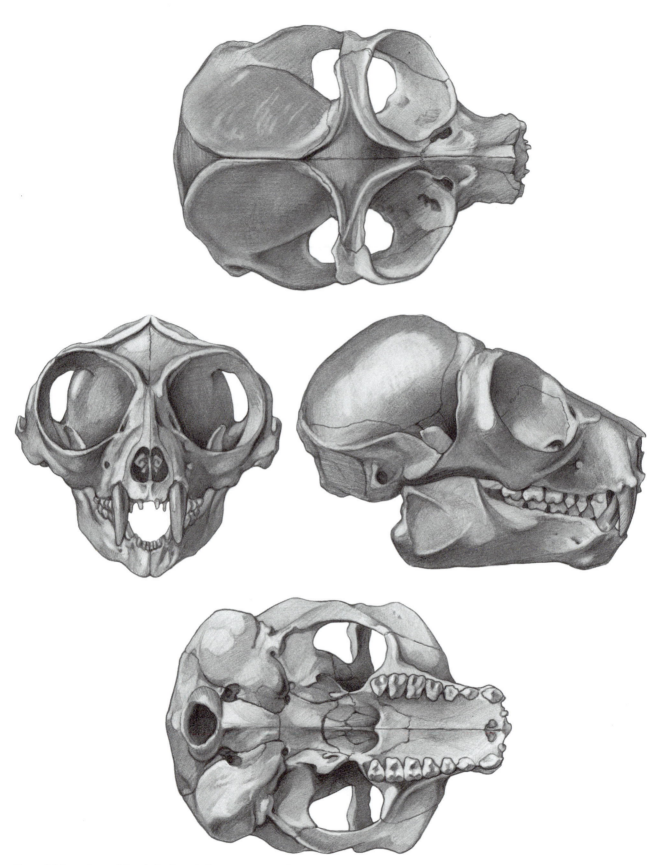

Figure 5.9 Four views of the skull of *Nycticebus coucang* (top, front, side, and below).

Orbit size in living primates has been correlated with day or night activity cycles. Enlarged orbits occur among nocturnal primates, while relatively smaller orbits are found in diurnal, or daytime, primates. This morphology to behavior correlation among living primates allows us to infer activity cycles for fossil primates on the basis of relative orbit size. Two Eocene fossil primates, *Shoshonius cooperi* (a), a small haplorhine North American omomyid primate, and *Adapis parisiensis* (b), a lemur-sized strepsirhine European adapiform, exhibit dramatically different orbit sizes relative to the size of their faces. The large orbits in *Shoshonius* suggest nocturnality, while the small orbits of *Adapis* imply diurnality.

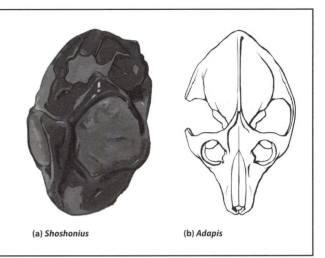

(a) *Shoshonius*   **(b) *Adapis***

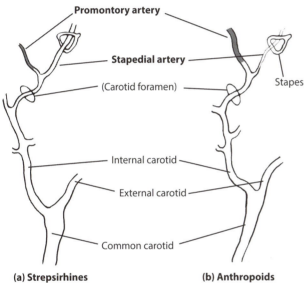

(a) Strepsirhines   **(b) Anthropoids**

Figure 5.10 Carotid artery circulation in primates: (a) shows the strepsirhine pattern, with a large stapedial artery threading through the stapes bone of the inner ear, while (b) illustrates the carotid pattern among anthropoids, where the stapedial artery is largely absent.

In terms of overall shape, the strepsirhine braincase is generally smooth although many taxa possess a sagittal crest at the top and at the posterior edge of the skull for the attachment of temporalis muscles. The temporalis muscle attaches to the sagittal crest and is an important chewing muscle, particularly for the anterior dentition (see fig. 5.5). Strepsirhine braincases also possess cresting posteriorly on the occipital bone for the muscles of the neck (i.e., the nuchal region). Underneath the skull, the foramen magnum, the opening where the spinal cord enters the brain-

stem and brain, is well posterior in strepsirhines (see fig. 5.1). Another distinctive strepsirhine skull feature is that the frontal bone is unfused with a metopic suture running down the middle of the frontal bone's length (see figs. 5.1, 5.7, 5.8, and 5.9).

Oddities

The skull of the aye-aye (*Daubentonia*) is far more compact and rounded, with a shorter nose and a relatively large brain compared to other Malagasy lemurs (fig. 5.11). Lorises have orbits and eyes that face upward, being less frontated, relative to other strepsirhine skulls (compare figs. 5.8 and 5.9).

HAPLORHINE HEADS

The skull of a tarsier, as a haplorhine primate, is different from that of living strepsirhines. First, the orbits are huge and the nose is short (fig. 5.12; see also fig. 5.2). Second, the internal nasal cavity of tarsiers has lost several of the turbinate bones; the remaining bones are much smaller relative to those of strepsirhine nasal chambers (fig. 5.12). The trend to reduce nose length and turbinates is also found among anthropoid primates. Third, both tarsiers and anthropoids possess a vertically oriented nasolacrimal duct relative to the horizontal orientation of this duct within the long-nosed strepsirhines.

Like the orbits, the braincase is relatively large in tarsiers as it is in anthropoids (figs. 5.12, 5.13, and 5.14). The back of the tarsier orbit has received a

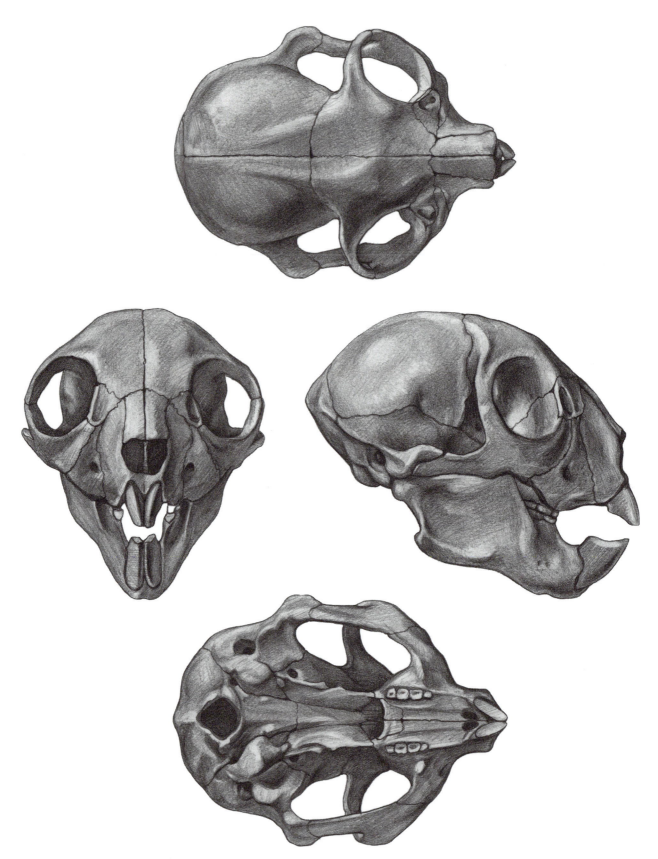

Figure 5.11 Four views of the skull of *Daubentonia madagascariensis* (top, front, side, and below).

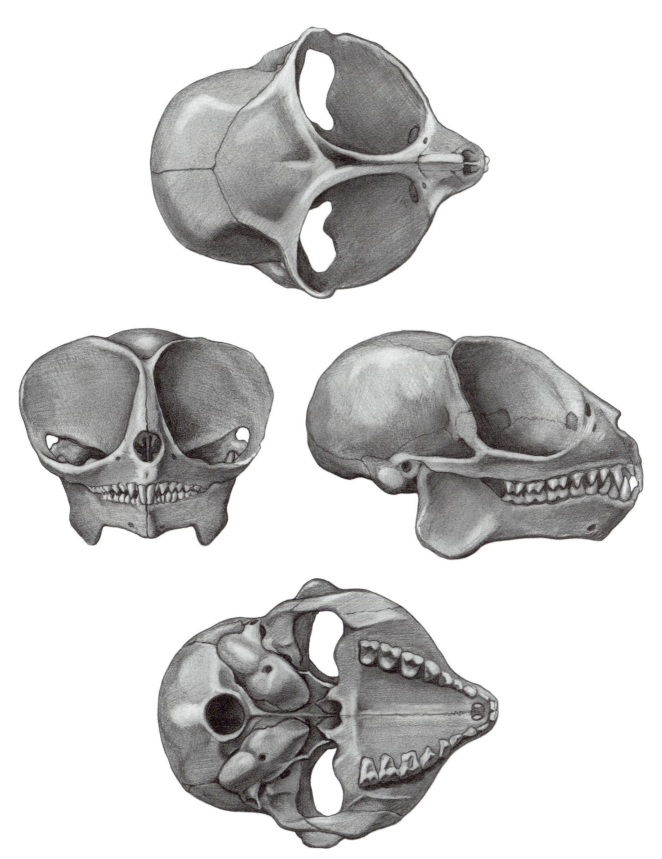

Figure 5.12 Four views of the skull of *Tarsius syrichta* (top, front, side, and below).

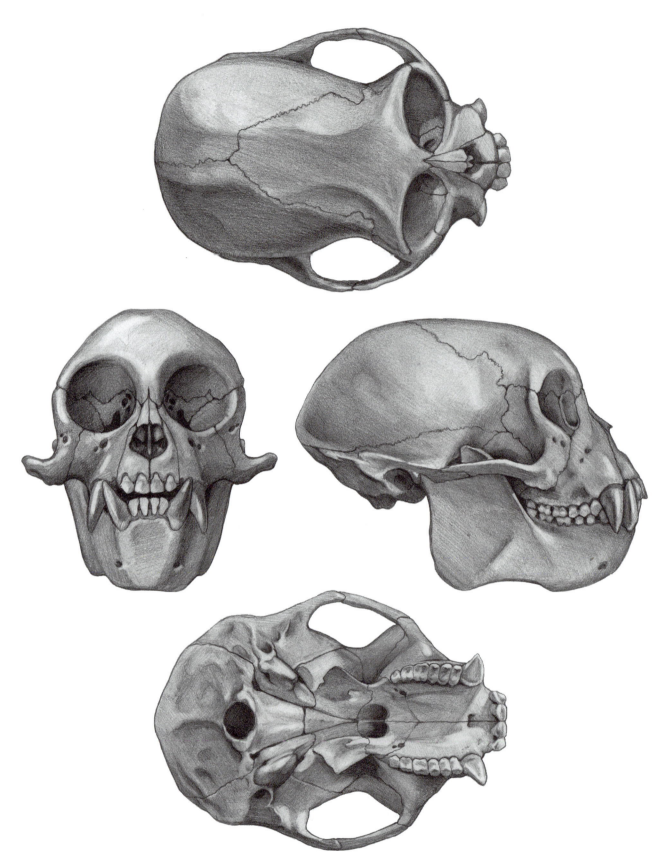

Figure 5.13 Four views of the skull of *Cebus capucinus* (top, front, side, and below).

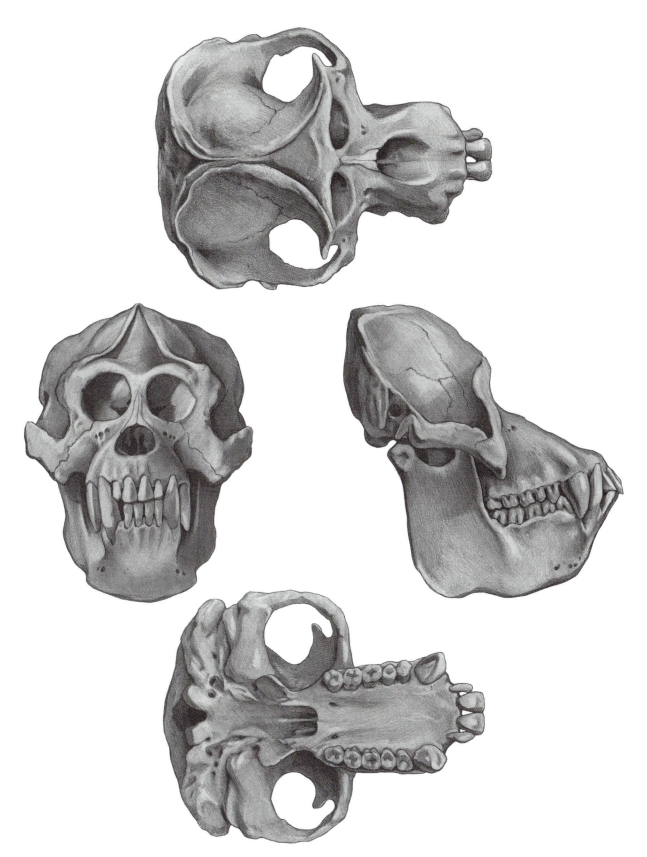

Figure 5.14 Four views of the skull of *Pongo pygmaeus* (top, front, side, and below).

lot of research attention since it is has long flanges projecting around its back. These flanges, parts of the zygomatic bones, the frontal bones, and the alisphenoid, form the partial postorbital closure condition in tarsiers, with spatial openings below and to the sides of the orbit. These postorbital bony extensions are not fused together, however, relative to anthropoid skulls. Complete postorbital closure is a key cranial adaptation found among all anthropoid skulls (see fig. 5.5).

In tarsiers, the auditory bullae are large and the internal carotid enters at the center of the bulla, a different entry point relative to that of the wet-nosed primates (see fig. 5.6). Another anatomical distinction is the ectotympanic ring, which extends outward into a tube for tarsiers and is commonly called a tubular ectotympanic (see fig. 5.6). Like galagos, tarsier heads have tall and mobile ears in terms of soft tissues to help them orient and hunt live prey, while the ears of anthropoids are relatively smaller.

When we compare the other haplorhine group, the anthropoids, in terms of head shape, we immediately see that anthropoid skulls have much larger braincases, and most possess greatly shortened noses (figs. 5.13 and 5.14), except for some Old World monkeys, relative to strepsirhine head shapes (see fig. 5.8). Anthropoids all possess a wall of bone behind their orbits closing off the orbits from the braincase (postorbital closure). This is a distinct anatomical feature for this group of primates and an easy way to identify anthropoid skulls from those of strepsirhines and tarsiers. This bony wall of closure separates the orbit from the cheek and temporal areas as it isolates the eye from the chewing muscles that attach behind the orbit. In terms of orbital orientation, anthropoid eyes face completely forward (orbital convergence) and are aligned vertically (high orbital frontation) relative to those of other primates. The frontal bone is fused in anthropoids (see fig. 5.7) in contrast to the two-bone condition in strepsirhines and tarsiers.

In terms of arterial blood supply to the brain, the stapedial branch of the internal carotid artery is lost in anthropoids and the promontory branch is far larger (see fig. 5.10). The promonotory branch represents the main carotid supplier of arterial blood to the relatively larger brains of anthropoids, along with the vertebral arteries. Mechanically, a single large-diameter

tube, the increased size of the promontory artery in anthropoids, actually provides greater blood volume to the brain relative to two smaller tubes (i.e., the promontory and stapedial arteries in strepsirhines). The internal carotid enters the auditory bulla from a medial or central location in anthropoids in contrast to the posterolateral entry in strepsirhines or the center entry for tarsiers (see fig. 5.6).

When we look closely at anthropoid skull anatomy, several features can be identified that allow us to separate these skulls into different anthropoid groups. For example, in South American monkeys, the ectotympanic bone is a ring that attaches to the outside rim of the petrosal, while in Old World monkeys and apes this ring is extended into a tubular condition that is longer in these taxa, a condition also found in tarsier skulls.

In a similar manner, South American monkey skulls (platyrrhines) can be sorted from those of Old World monkeys and apes (catarrhines) by the articulation of

The 14- to 8-million-year-old fossil ape named *Sivapithecus* displays a facial morphology similar to that of the living orangutan (*Pongo*)—although both possess different dental adaptations. Comparative anatomists have utilized this similar facial morphology to suggest that *Sivapithecus* represents an ancestral link to the orangutan lineage. What do you think?

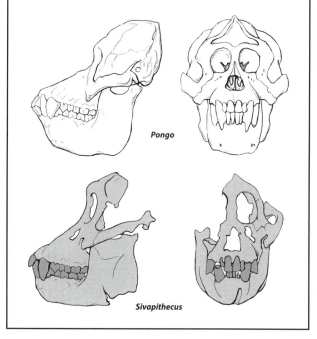

Pongo

Sivapithecus

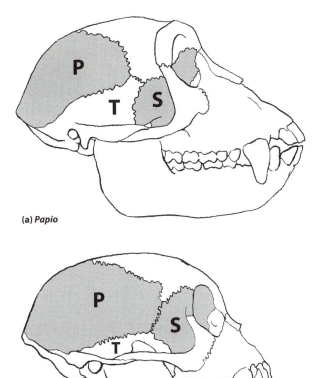

(a) Papio

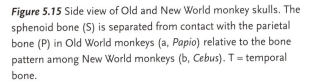

(b) Cebus

Figure 5.15 Side view of Old and New World monkey skulls. The sphenoid bone (S) is separated from contact with the parietal bone (P) in Old World monkeys (a, *Papio*) relative to the bone pattern among New World monkeys (b, *Cebus*). T = temporal bone.

bones along the side of the head (fig. 5.15). In platyrrhines, the parietal bone of the skull contacts the zygomatic bone, keeping the frontal and sphenoid bones separated along the side of the skull. In contrast, the parietal bone in catarrhines does not articulate with the zygomatic. The frontal and sphenoid bone articulation blocks parietal contact with the zygomatic in catarrhines.

Another skull feature has to do with air pockets. Anthropoids have pneumatized petrosal bones, which means that these bones contain many air pockets. The sphenoid sinus is common to all apes, while the maxillary, ethmoidal, and frontal sinuses are better developed in some.

BRAINS

When we compare the brains of primates, they tend to reflect the adaptive abilities of each group. For example, most strepsirhine brains are relatively small with a rather smooth neocortex without many folds or gyri, while the olfactory bulbs (cranial nerve I from the brain involved with smelling) are relatively large (fig. 5.16). Folding increases the surface area of the brain and some taxa like *Eulemur* or *Nycticebus* have more gyri relative to other strepsirhines. In contrast to the strepsirhines, anthropoid brains are relatively large with many fissures (fig. 5.16), except for the smooth-brained callitrichines. The olfactory bulbs of anthropoids are small, emphasizing the anthropoid adaptive shift away from smelling and toward vision. Tarsier brains are in between since they possess a smooth neocortex, like that of strepsirhines, but with a reduced olfactory region similar to that of anthropoids. Even the small-sized and smooth-brained callitrichines possess a brain roughly three times the size of a comparably sized strepsirhine primate, indicating the great increase in brain size across anthropoid primates. In general, primate brains increase in overall size with anthropoid phylogeny, they reduce olfaction relative to vision, and they increase tactile sensation.

Primate brain anatomy reflects these adaptive shifts. The sylvian fissure, a lateral brain sulcus (groove or furrow), separates the frontal and temporal lobes of the brain; the calcarine fissure is a sulcus that divides the visual cortex into two in all primates. The central sulcus, a groove between the frontal and parietal lobes of the brain, separates motor function, anterior to this sulcus, from the sensory aspects of the brain, the parietal regions or posterior section relative to this sulcus, thereby dividing the brain into front and back aspects (fig. 5.16). This sulcus is not always found in primates, being absent in strepsirhines, tarsiers, and callitrichines. The cerebellum, the region of the brain involved with body movements and balance, also increases in relative size phylogenetically, being smallest in lemurs and largest in apes.

Primate brains have been selected for an increased size of the neocortex and this appears to be tied to information-processing capabilities. Although a gorilla brain is hundreds of times larger than that of a mouse

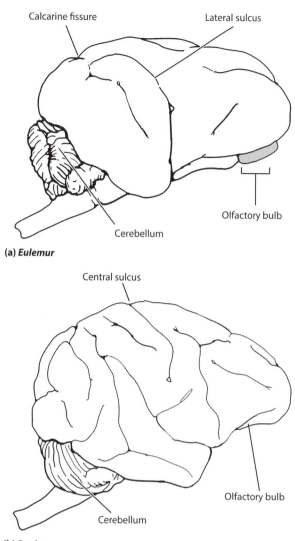

Calcarine fissure

Lateral sulcus

Olfactory bulb

Cerebellum

(a) *Eulemur*

Central sulcus

Cerebellum

Olfactory bulb

(b) *Papio*

Figure 5.16 Primate brain anatomy in *Eulemur* (a) and *Papio* (b).

lemur, both have a high density of cortical neurons relative to other mammalian brains, and haplorhine primates appear to possess a higher density of cortical neurons relative to strepsirhines. As one might suspect, much of primate neocortex enhancement is linked to the visual system.

There is, however, a considerable debate in the field concerning the strongest selective agent for an enhanced neocortex. Why did large brains evolve? This debate centers on two topics: ecology and sociality. For example, in terms of diet, fruit-eating primates possess relatively larger brain sizes in comparison to those of folivores. This is believed to select for cognitive abilities to exploit the environment since fruit is a patchy food source scattered throughout the forest

canopy. Frugivores, with their larger home ranges, thereby need to increase spatial memory to find this food source relative to the location of leaves eaten by folivores. The visual cue of discriminating color may also be linked to frugivory and neocortex expansion.

On the other hand, social group size has also been used as a good predictor of neocortical size among primates. Social cognition depends on the visual processing of rapid and complex social interactions. An enhanced neocortex has been linked with primate life in large groups where individuals must interpret information quickly among conspecifics. These studies suggest that brain size increase in primates is tied to the visual processing of social information.

EYES

Primates are visually oriented mammals and their eyes, and brains, reflect this sensory adaptation. Primate eyes (fig. 5.17), having shifted forward relative to those of other mammals, possess stereoscopic, or three-dimensional, vision. This overlap of the visual fields for each eye is greatest in anthropoids due to their shorter noses and their more convergent and highly frontated orbits. As the eyes migrated forward, the wiring of the brain also increased in size, especially for the occipital lobes of the neocortex. The increase in visual wiring added an extra set of optic nerve tracts among primates. All mammals possess contralateral, or crossing, nerve fibers that travel from one eye to the opposite side of the brain (e.g., the right eye nerve track via the optic nerve crosses to the left occipital lobe). Primates, like cats, have added a nerve pathway called the ipsilateral, or same side, nerve track, which travels back to the occipital lobe located on the same side as that of the eye. Thus, the visual field of the right eye is divided in half, with the inside half or the visual field closest to the nose traveling to the left occipital lobe via contralateral fibers while the visual field of the outside half of the right eye travels via ipsilateral fibers to the right occipital lobe (fig. 5.17). This means that every object a primate looks at has its image split in half by each eye, thereby generating four visual impressions that are projected back to the right and left occipital lobes of the brain. The right occipital lobe receives visual input from the right half of the visual field of the left eye (inside half)

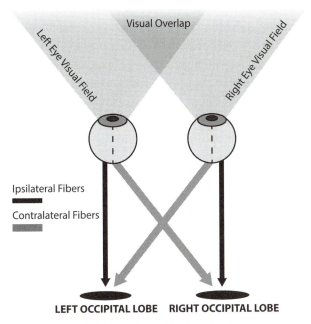

Figure 5.17 Visual field overlap and brain wiring in primates.

Eyes are dependent on light and thus nocturnal versus diurnal primate light sensitivity depends on the retina and its photoreceptors, the rods and cones. Rods dominate in the eyes of nocturnal species while cones are most common in eyes of diurnal anthropoids. Rods are best in low light conditions, while cones are adapted for bright light and visual acuity. The low light condition requires sensitive photoreceptors given the low levels of incoming light; because they are less able to discriminate, several rods connect to a single ganglion (a nerve cell). In contrast, each cone has only a single ganglion connection for greater acuity. Cones have greater discrimination but are less sensitive than rods. Rods, as photoreceptors, are tuned to a single wavelength and are achromatic. Color provides little help in distinguishing shapes at night (e.g., insect body shapes).

Color vision in anthropoids is more complicated than vision among strepsirhines or tarsiers. All anthropoids have color vision but some are trichromatic (three-color viewers) and others are dichromatic (two-color viewers). Catarrhine primates and some platyrrhines (Cebidae and Atelidae, female callitrichines) have trichromatic color vision. Male callitrichines are dichromatic, as they possess only a single X chromosome. Trichromatic vision simply means that these primates possess three retinal cones for specific blue, green, and red wavelengths and that they are able to make spectral discrimination for the different hues of these three colors, especially the red-green hues. In contrast, dichromatics see yellow, brown, green, and red as the same color. The gene for the short wavelength, blue, is located on an autosomal chromosome, while the middle (green) and long (red) wavelengths are located on the X chromosome. Thus, given the XX (female) or XY (male) chromosome pattern in mammals, the medium and long wavelengths exhibit greater variation in color perception. Trichromatic color vision in anthropoids has been suggested to allow for the discrimination of young, often red leaves, as well as fruits, among a green background in the canopy. Young leaves lack toxins that mature leaves build up and therefore are more digestible by primates.

One way to increase light absorption in a dark environment is to add crystals to the retina. The eyeshine of primates, usually yellow to yellow-orange

and the right half of the visual field (outside half) from the right eye. The left occipital lobe receives visual input from the left half of the visual field of the right eye and the left half of the visual field from the left eye. The brain, specifically the right and left occipital lobes, then recombines all of these retinal images into what is actually perceived as a visual object.

One of the key advantages of moving the orbits forward is the realignment of the visual and optic axes of the eye. This realignment, which brings the two axes closer together at the midpoint of the eyeball, allows light to pass through the center of the lens, which is less curved. The center of the lens is therefore optimal for light to pass through since its lack of curvature reduces spherical aberration or blurred images. By reducing pupil size, primates allow light to pass through the center of the lens and this realignment contributes to improved image quality. The forward movement of the primate orbits for stereoscopic vision enhances light sensitivity, allows for better optical discrimination, and expands depth perception. Nocturnal primates need to capture as much light as is possible to see at night but their inability to reduce pupil size affects image quality. Closer alignment of the optic and visual axes in their eyes allows nocturnal primates to view items without disrupting their light-gathering abilities.

in coloration, comes from the tapetum lucidum, a reflecting zone in the back of the eye between the choroid and retina that serves to increase sensitivity in low light conditions. This structure has flat crystals of riboflavin, vitamin B-2, which reflect light back through the retina, shifting the wavelengths of light toward optimum sensitivity for the light receptors. Many nocturnal mammals possess this special structure, as do strepsirhine primates. A tapetum lucidum is lacking in all tarsiers and anthropoids but appears in almost all species of strepsirhines, both nocturnal or diurnal.

In haplorhine primates, the retina has a depressed region or a pit called the fovea. The fovea and the macula lutea, the area immediately surrounding the fovea that is called the yellow spot, are part of the central visual field and both make up an avascular area of the retina. Cones are concentrated in this region, especially in anthropoids. The fovea in anthropoids is therefore an area of enhanced visual acuity. In tarsiers, the fovea and macular lutea are also composed of cones. Why does this nocturnal species have a concentration of retinal cones and a retinal fovea? The simplest answer is an evolutionary one: the ancestral haplorhine, and thereby ancestor to tarsiers, was derived from a diurnal predecessor. An opposite adaptive pattern shows a fovea and a macula lutea as absent in the nocturnal owl monkey, *Aotus*, and suggests a secondarily nocturnal condition for *Aotus*.

SMELL

The sense of smell involves two regions of the nose, the rhinarium, if present, and the olfactory epithelium of the nasal passages. The rhinarium is a wet and hairless area surrounding the nostrils and is present only in strepsirhine primates and most other mammals (fig. 5.18). It is absent in haplorhines, tarsiers, and anthropoids, representing a derived condition (fig. 5.18). In wet-nosed primates, the rhinarium is connected to an olfactory structure called the vomeronasal organ, or Jacobson's organ, by the median groove. This organ has its own nerve to the accessory olfactory bulb and seems to be able to recognize species-specific odors like those connected with reproductive state.

Smelling is correlated to the number of ethmotur-

binals and the olfactory epithelium lining the surfaces of the nasal cavity. These surfaces send nerve fibers through the cribiform plate, part of the ethmoid bone, to the olfactory bulb (cranial nerve I). The number of turbinals increases the surface area for the olfactory epithelium and therefore long-nosed primates like strepsirhines have a better olfactory sense with larger olfactory bulbs compared to tarsiers or anthropoids, primates with reduced turbinals, nose length, and olfactory bulb size.

JAWS

Strepsirhine and tarsier mandibles are long and narrow in height, and they possess three processes proximally: the coronoid process, the condyloid process (i.e., the actual joint surface that articulates with the skull at the temporomandibular joint), and the angular process (fig. 5.19). Strepsirhine and tarsier jaws are unfused anteriorly at the symphysis, making two separate bones, or the right and left halves of the mandible. The tooth row is functionally aligned with the condyloid process, or the joint condyle; this middle mandibular process is lower than the coronoid process in strepsirhines and tarsiers (fig. 5.19). In contrast, anthropoid mandibles have lost the angular process; they are taller, particularly along the ascending ramus, the condyloid process reaches a similar height to the coronoid process, and the two mandibular halves are fused anteriorly at the symphysis (fig. 5.19). Both jaw types are distinctive across primates, and their structural differences reflect dietary preferences.

When the condyloid process, or the hinge of the jaw, is low relative to the coronoid process, the force "in lever" is long and muscular force is increased for the temporalis muscle. In contrast, when the condyloid process is approximately equal in height with the coronoid process (fig. 5.19), the force "out lever" is increased for masseter function (fig. 5.20). Thus, the changing height of the condyloid process modifies the lever mechanics of the jaw during chewing. Insect-eating primates possess low condyloid processes, while frugivores and folivores tend to have more highly positioned condyloid processes for masseter and pterygoid muscle function.

The upper dental row of primates is made up of

two bones, the premaxilla and the maxilla (see fig. 5.1). All incisors lie anterior to the premaxillary-maxillary suture, making this suture the demarcation line that separates upper incisors from canines. Most upper teeth have roots that extend into the maxilla.

Like other mammals, primates chew, in contrast to the rip, tear, and swallow method of feeding found in crocodiles and other animals. Primate jaws do not simply open and close but move in a figure-eight motion and function as a class-three lever (e.g., like a door; see chapter 4). In class-three levers the fulcrum or hinge is far to one side. The mandibular condyle is the fulcrum and the temporomandibular joint the hinge in this lever analogy. In class-three levers, the resistance arm is always long. It therefore takes a greater muscular effort to overcome the resistance of the jaw (i.e., gravity) when closing the jaw for chewing, but a small muscular effort will produce a large amount of motion. The advantage of this lever system is speed

and range of motion at the expense of force. Chewing muscles therefore pull on the jaw to close the mouth and to compress food (the resistance force).

How does all of this happen? When a primate opens its mouth, muscles pull inward on the mandibular condyles and the symphysis in the front of the mandible is deformed. When closing the mouth, one condyle is pulled outward and chewing occurs on one side of the mandible (unilateral chewing), leaving only the biting side occluded. Closing the mouth and biting produces the power stroke of bite force compression when primates chew food. The acts of biting and chewing cause dorsoventral shearing and twisting of the mandible anteriorly along the incisor region. As primates eat tougher foods, greater muscular force is required for biting and chewing. As chewing forces increase, jaws become thicker and taller vertically, and the mandible is fused at the symphysis to counter these bony stresses.

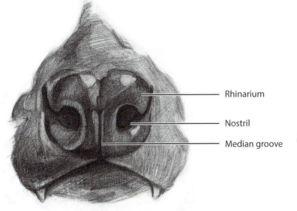

(a) *Eulemur*

Rhinarium

Nostril

Median groove

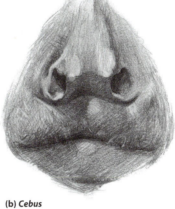

(b) *Cebus*

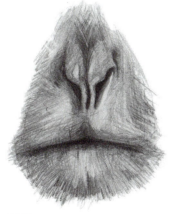

(c) *Macaca*

(d) *Gorilla*

Figure 5.18 Primate nose anatomy.

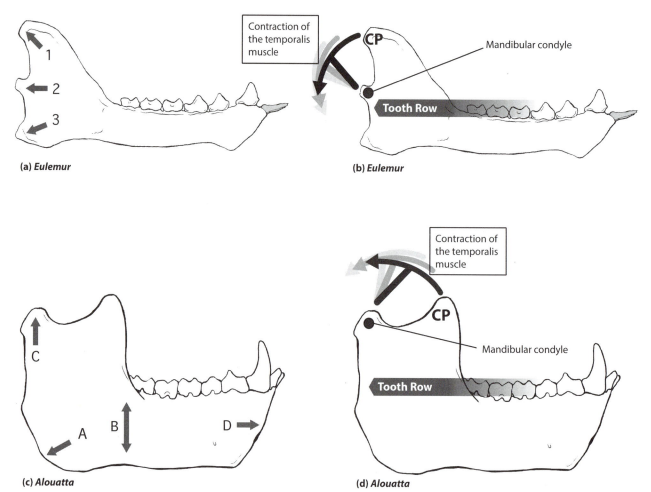

(a) *Eulemur*

(b) *Eulemur*

(c) *Alouatta*

(d) *Alouatta*

Figure 5.19 Two types of primate mandibles. *Eulemur* (a, b) illustrates a strepsirhine jaw where the tooth row is aligned with the mandibular condyle. It has (1) a coronoid process, (2) the mandibular condyle, and (3) an angular process. In an anthropoid mandible like that of *Alouatta* (c, d), the symphysis is fused (D), the mandibular body is tall (B), the angular process is absent (A), and the actual jaw joint, the condylar process (C), has shifted upward, making its location about equal in height with the coronoid process.

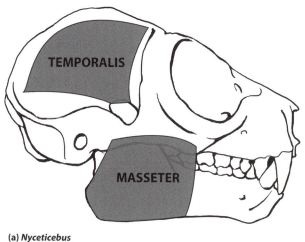

(a) *Nycticebus*

Figure 5.20 Important chewing muscles for primates.

The oral tract and tongue of primates are similar to that of other mammals. Primate tongues are long and typically possess papillae, or bumps. Most taste buds are located within the largest papillae, but some taste receptors line the sides and back of the tongue. Under the tongue in strepsirhines there is a sublingual structure that helps clean between the closely apposed teeth of the toothcomb. The sublingua, a keratinized structure in strepsirhines, has a median ridge and is split into two halves with pointy edges along the periphery. The aye-aye, which lacks a toothcomb, has a simplified sublingua, as do tarsiers. Anthropoids generally lack a sublingua, while callitrichids possess a reduced structure.

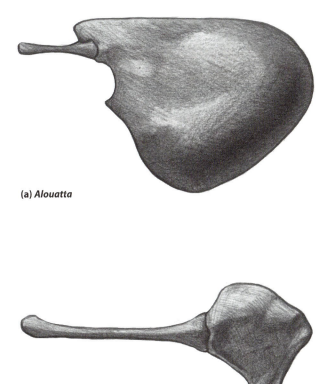

(a) *Alouatta*

(b) *Ateles*

Figure 5.21 The primate hyoid bone. This bone is significantly enlarged in *Alouatta* (a) relative to the more normal and smaller bone condition exhibited by *Ateles* (b), a close relative of *Alouatta*.

ODDITIES

Enlarged larynxes or laryngeal sacs are noted in many primates where loud, long-distance vocalizations are important social communications. For example, *Alouatta* (fig. 5.21) has an enlarged hyoid bone that it uses as a resonating chamber for its loud, booming calls. Likewise, siamangs possess a large, descendible, and inflatable laryngeal throat sac for use in calling, and male orangutans possess large laryngeal sacs to help boom out their long call.

In terms of skull anatomy, the owl monkey, *Aotus*, the only nocturnal anthropoid, has large orbits and eyes. Several species of Old World monkeys, especially the baboons, have long muzzles,. In this anthropoid case, however, it is the roots of their large teeth and not their olfactory sense that have increased rostral lengths. Orangutans are known for their highly sloping or dished facial profiles and oblong-shaped orbits relative to other primate faces.

Selected References

Allman, J. 1977. Evolution of the visual system in the early primates; pp. 1–53 *in* H.W. Reese, M.A. Epstein, and A.N. Epstein (eds.), Progress in Psychobiology, Physiology, and Psychology, Vol. 7. Academic Press, New York.

Ankel-Simons, F. 2000. Primate Anatomy—An Introduction. 2nd edition. Academic Press, New York.

Armstrong, E., and D. Falk (eds.). 1982. Primate Brain Evolution: Methods and Concepts. Plenum Press, New York.

Barton, R.A. 1998. Visual specialization and brain evolution in primates. Proceedings of the Royal Society of London 265:1933–1937.

———. 2006. Primate brain evolution: integrating comparative, neurophysiological, and ethological data. Evolutionary Anthropology 15:224–236.

Cartmill, M. 1975. Strepsirhine basicranial structures and affinities of Cheirogaleidae; pp. 313–354 *in* W.P. Luckett (ed.), Phylogeny of Primates. Plenum Press, New York.

Dominy, N.J., P.W. Lucas, D. Osorio, and N. Yamashita. 2001. The sensory ecology of primate food perception. Evolutionary Anthropology 10:171–186.

Elbroch, M. 2006. Animal Skulls—A Guide to North American Species. Stackpole Books, Mechanicsburg, Pa.

Fleagle, J.G. 1999. Primate Adaptation and Evolution. Academic Press, New York.

Fleagle, J.G., and R.F. Kay (eds.). 2004. Anthropoid Origins. Plenum Press, New York.

Hartman, C.G., and W.L. Straus (eds.). 1971. The Anatomy of the Rhesus Monkey. Hafner Publishing Company, New York.

Heesy, C.P. 2009. Seeing in stereo: the ecology and evolution of primate binocular vision and stereopsis. Evolutionary Anthropology 18:21–35.

Hershkovitz, P. 1977. Living New World Primates (Platyrrhini), with an Introduction to Primates. University of Chicago Press, Chicago.

Hiiemae, K.M. 1978. Mammalian mastication, a review of the activity of the jaw muscles and the movements they produce in chewing; pp. 359–398 *in* P.M. Butler and K.A. Joysey (eds.), Development, Function and Evolution. Academic Press, New York.

Hylander, W.L. 1984. Stress and strain in the mandibular symphysis of primates: a test of competing hypotheses. American Journal of Physical Anthropology 64:1–46.

Hylander, W.L., K.R. Johnson, and A.W. Crompton. 1987. Loading patterns and jaw movements during mastication in *Macaca fascicularis*: a bone-strain, electromyographic, and cineradiographic analysis. American Journal of Physical Anthropology 72:287–314.

Jungers, W.L. (ed.). 1985. Size and Scaling in Primate Biology. Plenum Press, New York.

Kay, R.F., and K.M. Hiiemae. 1974. Mastication in *Galago crassicaudatus*: a cinefluorographic and occlusal study; pp. 501–

530 *in* R.D. Martin, G.A. Doyle, and A.C. Walker (eds.), Prosimian Biology. Duckworth Publishing, London.

Le Gros Clark, W.E. 1959. The Antecedents of Man: An Introduction to the Evolution of Primates. University of Edinburgh Press, Edinburgh.

Martin, R.D. 1990. Primate Origins and Evolution—A Phylogenetic Reconstruction. Princeton University Press, Princeton, N.J.

Pettigrew, J.D. 1986. The evolution of binocular vision; pp. 208–222 *in* J.D. Pettigrew, K.J. Sanderson, and W.R. Levick (eds.), Visual Neuroscience. Cambridge University Press, Cambridge, UK.

Ravosa, M.J., V.E. Noble, W.L. Hylander, K.R. Johnson, and E.M. Kowalski. 2000. Masticatory stress, orbital orientation and the evolution of the primate postorbital bar. Journal of Human Evolution 38:667–693.

Rosenberger, A.L. 1986. Platyrrhines, catarrhines and the anthropoid transition; pp. 66–88 *in* B. Wood, L. Martin, and P. Andrews (eds.), Major Topics in Primate and Human Evolution. Cambridge University Press, Cambridge, UK.

Ross, C.F. 1995. Allometric and functional influences on primate orbit orientation and the origins of the Anthropoidea. Journal of Human Evolution 29:201–227.

Ross, C.F., and R.F. Kay (eds.). 2004. Anthropoid Origins—New Visions. Kluwer Academic–Plenum Publishers, New York.

Ross, C.F., and E.C. Kirk. 2007. Evolution of eye size and shape in primates. Journal of Human Evolution 52:294–313.

Rossie, J.B. 2005. Anatomy of the nasal cavity and paranasal sinuses in *Aegyptopithecus* and Early Miocene African catarrhines. American Journal of Physical Anthropology 126:250–267.

Rossie, J.B., X. Ni, and C. Beard. 2006. Cranial remains of an Eocene tarsier. Proceedings of the National Academy of Sciences 103:4381–4385.

Simons, E.L., and D.T. Rasmussen. 1989. Cranial morphology of *Tarsius* and *Aegyptopithecus* and the question of the tarsier-anthropoidean clade. American Journal of Physical Anthropology 19:1–23.

Smith, T.D., J.B. Rossie, and K.P. Bhatnagar. 2007. Evolution of the nose and nasal skeleton in primates. Evolutionary Anthropology 16:132–146.

Swindler, D.R. 1998. Introduction to the Primates. University of Washington Press, Seattle.

Wright, P.C., E.L. Simons, and S. Gursky (eds.). 2003. Tarsiers—Past, Present and Future. Rutgers University Press, New Brunswick, N.J.

6 Teeth

PRIMATE TEETH AND DIETS

Teeth represent the start of the food-processing component of digestion and energy uptake among primates. Primate teeth occur in distinctive patterns and numbers (table 6.1), but they are all designed to grab, crush, and partition food items into smaller units. All primates, like other mammals, have two sets of teeth (deciduous, or baby, teeth and a later adult dentition) and all teeth are composed of the same materials (enamel, dentine, and a pulp or nerve chamber) (fig. 6.1). Like other mammals, primates possess four types of teeth: incisors, canines, premolars, and molars (fig. 6.2).

Upper and lower incisors vary by number as well as breadth across primates and are generally used to grab food items and reduce them in size by nipping, slicing, or puncturing. Primate incisors are variable in shape—from tall, pointed teeth in taxa like tarsiers to the rodent-like, ever-growing incisors of the aye-aye (figs. 6.3 and 6.4). Across primate taxonomic groups, all living strepsirhines, with the exception of the aye-aye, have thin, blade-like lower incisors and canines that extend away from the mandible into a structure called a toothcomb (fig. 6.4). With the exception of the aye-aye, the other living strepsirhines have a large gap between their upper central incisors, which are generally tiny (fig. 6.5). This lower toothcomb structure with the gap above is a distinctive anterior dentition for strepsirhine primates and is primarily used by

lemurs, lorises, and galagos to scrape gums and groom fur. One odd incisor dental pattern in strepsirhines is found in the genus *Lepilemur* (fig. 6.6), a taxon that has lost its upper incisors altogether. For the living anthropoids, all monkeys and apes have spatula-shaped incisors (fig. 6.7). Anthropoids, like the pitheciines, have odd, peg-like lower incisors and procumbent upper incisors (fig. 6.3), while the marmosets (*Cebuella* and *Callithrix*) possess tall lower incisors equal in height to their canines.

In terms of diet, primate gumnivores (gum and sap feeders) require stout incisors, while primate frugivores tend to have broad upper incisors relative to the narrower incisors of folivores. The pointy incisors of tarsiers are clearly for insect and vertebrate capture and puncture. The rodent-like incisors of aye-ayes are for opening up holes within woody branches for their finger foray searches for hidden grubs.

Primate canines are less variable in shape and number (table 6.1). Most are tall and pointed or conical in shape, although several primates have either thicker or stout canine shapes (fig. 6.3) or slender, blade-like canines (fig. 6.8). Many anthropoid canines are sharpened on the posterior side, making them dagger-like, and this dagger shape allows a puncture and slice upon impact. Among the living strepsirhines, the lower canines are the outside teeth of the toothcomb structure but are lost in the odd-toothed *Daubentonia* (fig. 6.4). The South American pitheciines have laterally splayed and robust canines (fig. 6.3) in contrast to the long,

Table 6.1 Primate Tooth Formulas

Primate Families	Upper Dentition	Lower Dentition
Strepsirhini		
Lorisidae	2–1–3–3	2–1–3–3
Galagidae	2–1–3–3	2–1–3–3
Cheirogaleidae	2–1–3–3	2–1–3–3
Lemuridae	2–1–3–3	2–1–3–3
Lepilemuridae	0–1–3–3	2–1–3–3
Indriidae	2–1–3–3	1–1–2–3
Daubentoniidae	1–0–1–3	1–0–0–3
Haplorhini		
Tarsiidae	2–1–3–3	1–1–3–3
Callitrichidae	2–1–3–2	2–1–3–2
Callimico	2–1–3–3	2–1–3–3
Cebidae	2–1–3–3	2–1–3–3
Atelidae	2–1–3–3	2–1–3–3
Cercopithecidae	2–1–2–3	2–1–2–3
Hylobatidae	2–1–2–3	2–1–2–3
Pongidae	2–1–2–3	2–1–2–3

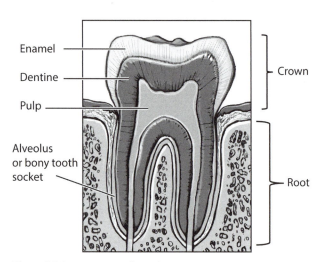

Figure 6.1 Cross section of a molar.

Enamel — Crown
Dentine
Pulp
Alveolus or bony tooth socket — Root

The number of teeth in living and fossil primate mandibles or maxillae can be counted, and these tooth assemblages are often distinctive for evolutionary lineages. Tooth formulas are counted by using half of the lower or upper dentitions. Starting at the midline, you can count the number of incisors, the single canine, premolars, and molars (see table 6.1 for living primate families). Illustrated are three jaws, (a) from an Eocene strepsirhine adapiform fossil primate from North America, *Notharctus tenebrosus*; (b) is the jaw of *Eulemur fulvus*, an extant strepsirhine lemurid from Madagascar; and (c) is the jaw of a modern human, *Homo sapiens*. As you can see, all three show two incisors, one canine, and three molars. The premolars, however, differ in number, with four for *Notharctus*, three for *Eulemur*, and only two for *Homo*. Several premolars have clearly been lost over evolutionary time. The ancestral tooth formula for primates is 2–1–4–3, as exhibited by *Notharctus*.

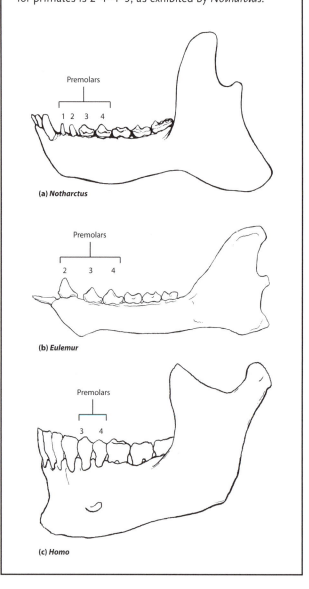

Premolars
1 2 3 4
(a) *Notharctus*

Premolars
2 3 4
(b) *Eulemur*

Premolars
3 4
(c) *Homo*

dagger-like canines of Old World monkeys (see fig. 6.8). For apes, stout and robust canines are the norm, especially in gorillas. Among sexually dimorphic primates, males have much taller and larger canines than do females (fig. 6.8). Canine size dimorphism is especially true in Old World monkeys and great apes.

Primate premolars are also variable in shape (fig. 6.9) with four premolars being the original number of premolars for several fossil primates. Most living primates have three premolars, while catarrhines possess only two. Premolar shape tends to correspond to dietary preferences with larger teeth, increased number of cusps, and cresting found in leaf-eating

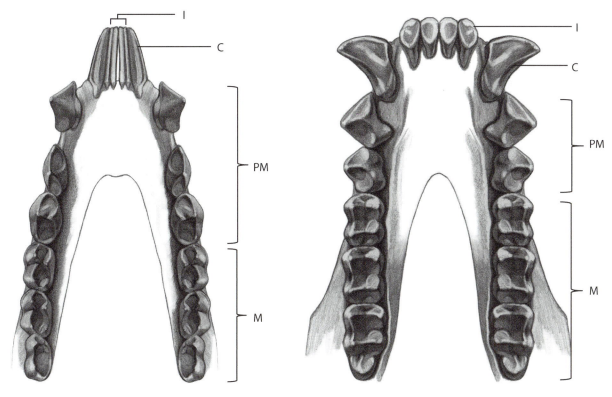

(a) *Eulemur*

(b) *Colobus*

Figure 6.2 Types of primate teeth illustrated in *Eulemur* (a) and *Colobus* (b). I = incisor; C = canine; PM = premolar; and M = molar. Note the toothcomb in *Eulemur*.

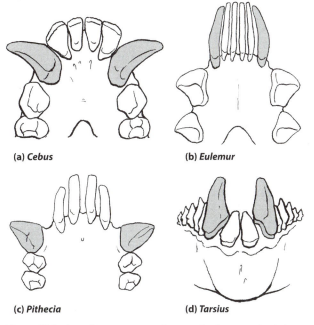

(a) *Cebus*

(b) *Eulemur*

(c) *Pithecia*

(d) *Tarsius*

Figure 6.3 Incisor shapes across primates. Canines are shaded.

primates and flatter tooth crowns in frugivores. Insect eaters possess tall and pointy premolars. In Old World monkeys, the first lower premolar (P3) is called a sectorial premolar; it has a long cutting edge for the upper canine to sharpen against (fig. 6.10). Apes have a smaller version of this P3 sectorial premolar, while platyrrhines use a lower P2, the first premolar after the canine.

Molars are more variable in shape, being tied to dietary preferences, but they only occasionally vary in number (table 6.1). Primate molars can possess tall cusps (e.g., insectivores), can have connected cusps or ridges (e.g., folivores), or can be flattened along their occlusal surface (e.g., fruit and seed eaters). Molar morphology is complicated in primates, and in mammals generally, and this anatomy has been used extensively to catalogue living and fossil primates. I explain the pattern of upper and lower molar cusp morphology below but in general it is simply a geometric pattern of triangles or rectangles with bumps.

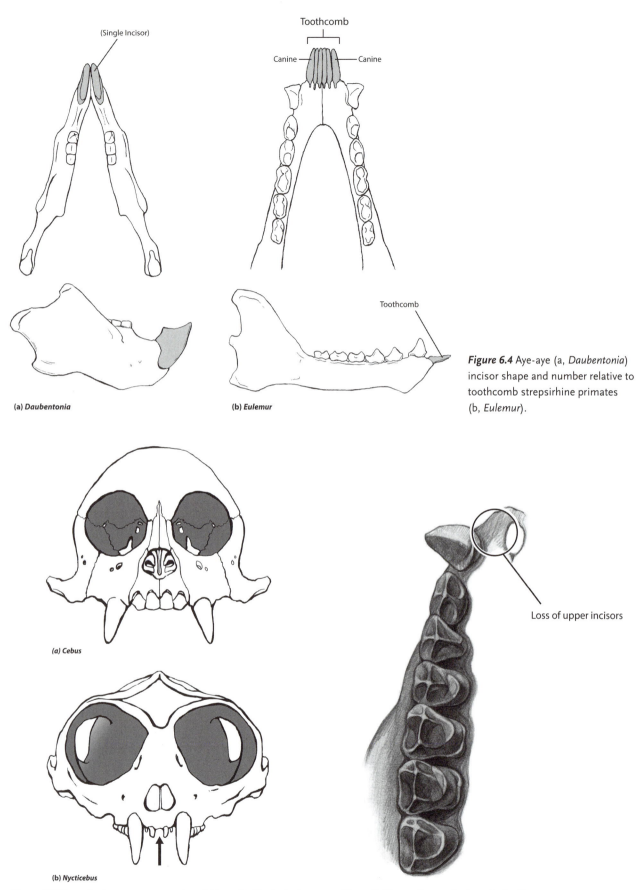

(Single Incisor)

Toothcomb

Canine —   — Canine

(a) *Daubentonia*

(b) *Eulemur*

Toothcomb

Figure 6.4 Aye-aye (a, *Daubentonia*) incisor shape and number relative to toothcomb strepsirhine primates (b, *Eulemur*).

(a) *Cebus*

(b) *Nycticebus*

Loss of upper incisors

Figure 6.5 Upper incisor gap among strepsirhines (b, *Nycticebus*) relative to anthropoid primates (a, *Cebus*).

Figure 6.6 Loss of upper incisors in *Lepilemur*.

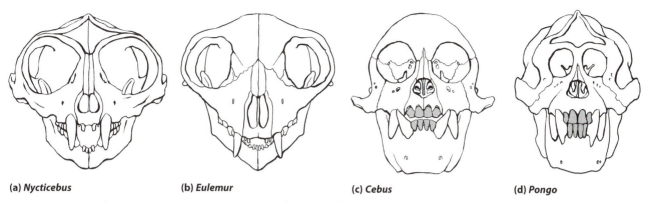

(a) Nycticebus   **(b) Eulemur**   **(c) Cebus**   **(d) Pongo**

Figure 6.7 Canine and incisor shapes across primates. Note the spatula-like incisor shapes in anthropoids (c, *Cebus*; d, *Pongo*).

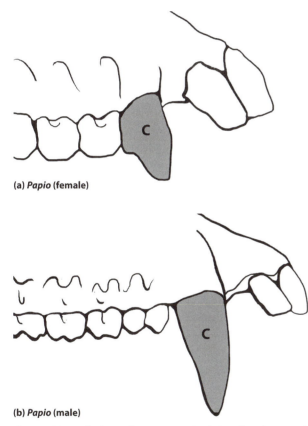

(a) Papio (female)

(b) Papio (male)

Figure 6.8 Sexually dimorphic canines (shaded) in female (a) and male (b) baboons (*Papio*).

MOLAR CUSPS

Primitively, primate upper molars possess three main cusps: the paracone, the metacone, and the protocone (fig. 6.11). This gives an upper molar an overall triangular shape. The basin between these three major cusps is the trigon. Sometimes smaller cusps lie in between the three large upper cusps; these smaller cusps are called conules (or specifically, the paraconule and the metaconule; fig. 6.12). Along the outside of the upper molar, extra enamel may form into a cingulum. Generally, the cingulum forms on the posterolingual, or tongue, side of the dentition in primates. Occasionally, it may form on the anterolingual side (the protostyle). If the external cingulum forms along the buccal side of a molar, we call this region the stylar shelf. The stylar shelf can be elaborated to form three styles: the parastyle, the mesostyle, or the metastyle (fig. 6.12). Perhaps more interesting in terms of taxonomy and phylogeny, the posterolingual cingulum of primates can develop into a fourth upper molar cusp, the hypocone, a common cusp among anthropoid upper molars (fig. 6.12).

All of the upper molar cusps in primates occlude with the lower molars, and vice versa. For example, the upper molar protocone occludes within the talonid basin of the lower molar (i.e., the lower molar trigonid must fit between two upper molars). Thus, cusp and basin shape functions as a single unit during mastication. Changing one part above requires changes below.

The primate lower molar often (or primitively) has five cusps: the paraconid, the metaconid, the entoconid, the protoconid, and the hypoconid (fig. 6.13). Note that all lower molar cusps are labeled -conids while cusps of upper molars are called -cones. The front region of the lower molar is the trigonid and this region includes three cusps: the paraconid, the metaconid (lingually), and the protoconid (buccally; fig. 6.13). The fourth and fifth lower molar cusps, the

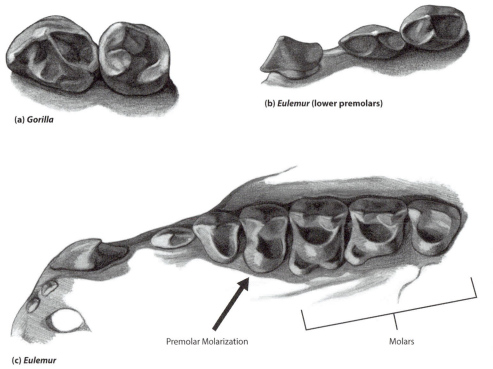

(a) Gorilla

(b) Eulemur (lower premolars)

Premolar Molarization

Molars

(c) Eulemur

Figure 6.9 Primate lower premolar shape variation in (a) *Gorilla* relative to (b) *Eulemur*. (c) illustrates the increased size of an upper premolar (premolar molarization) within the upper dentition of *Eulemur*.

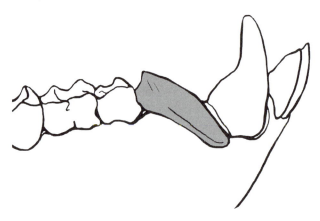

Figure 6.10 A sectorial lower premolar of a baboon (*Papio*).

entoconid (lingually) and the hypoconid (buccally), surround the talonid basin, the back basin of a lower molar. Occasionally, a sixth cusp, the hypoconulid, is added to this lower molar cusp configuration (fig. 6.13).

From upper and lower molar cusps, ridges or crests can connect different cusps together (figs. 6.11 and 6.13). We call upper molar crests cristas and use the label cristids for lower molar crests. Tall crests are found among the more folivorous primates to shear and cut plant material into smaller sizes for digestion.

Fruit eaters have small, hilly cusps, or bunodont molars. The seed-eating primates possess flat molars with thickened enamel and smaller and simpler cusp patterns. Insectivory requires tall, pointy, and sharp cusps to puncture and process the chitin (hard exoskeleton) of insect bodies.

SPECIALIZED DENTAL CHARACTERISTICS ACROSS PRIMATES

The teeth of primates are variable across their respective taxonomic families (table 6.2; figs. 6.14–6.19). Although *Lepilemur* has lost all of its upper incisors (see fig. 6.6), *Daubentonia* has lost most of its teeth altogether with a dental formula of 1–0–1–3/1–0–0–3. The aye-aye is clearly the primate winner for most unusual dental pattern (see below). Other primates have lost teeth as well (see table 6.1). Indriids have lost one lower incisor in their toothcomb (fig. 6.15), Old World monkeys and apes have lost a premolar (figs. 17–19), and callitrichids (except *Callimico*) have surprisingly lost their third molars (fig. 6.16).

In terms of general shape, the lower molar trigonid is higher than the talonid region in strepsirhines and tarsiers (fig. 6.20), relative to anthropoid primates,

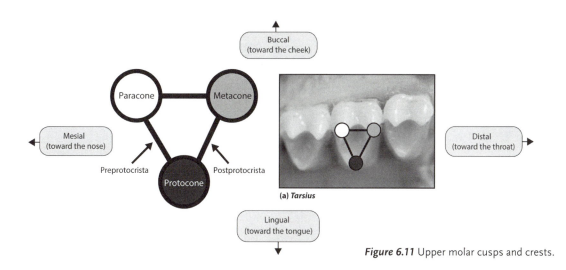

Buccal
(toward the cheek)

Paracone — Metacone

Mesial
(toward the nose)

Preprotocrista   Postprotocrista

Protocone

Distal
(toward the throat)

Lingual
(toward the tongue)

(a) *Tarsius*

Figure 6.11 Upper molar cusps and crests.

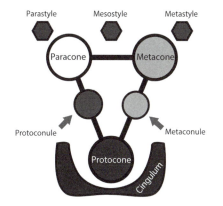

Parastyle   Mesostyle   Metastyle

Paracone   Metacone

Protoconule   Metaconule

Protocone

Cingulum

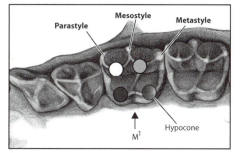

(a) *Propithecus* (upper molars + premolars)

Parastyle   Mesostyle   Metastyle

Hypocone

M¹

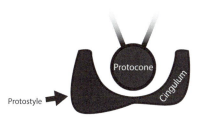

Protocone

Cingulum

Protostyle

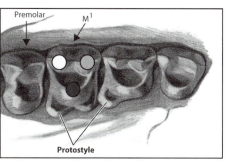

Premolar   M¹

Protostyle

(b) *Eulemur* (upper molars +premolar)

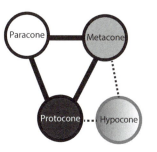

Paracone   Metacone

Protocone   Hypocone

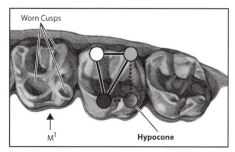

Worn Cusps

M¹   Hypocone

(c) *Gorilla* (upper molars)

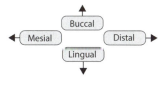

Buccal

Mesial   Distal

Lingual

Figure 6.12 Upper molar cusp modifications across primates. The addition of styles in *Propithecus* (a); a protostyle in *Eulemur* (b), and a hypocone in *Gorilla* (c).

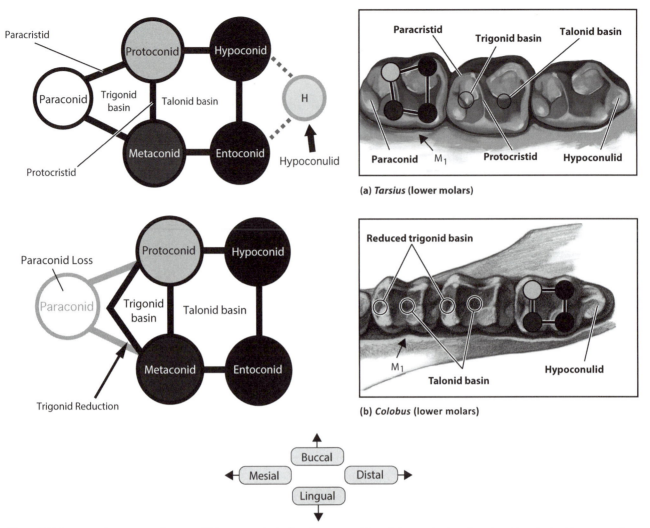

(a) *Tarsius* **(lower molars)**

(b) *Colobus* **(lower molars)**

Figure 6.13 Lower molar cusps and crests. All five cusps, including the paraconid, and the two basins, the trigonid and talonid, are seen in the lower molars of *Tarsius* (a), while (b) illustrates significant lower molar changes in anthropoids (*Colobus*). The trigonid is reduced in size and the paraconid is lost while the talonid basin expands.

where molars are more rectangular or quadrate in shape and the talonid basin is greatly expanded. Most living primates, except the tarsier, have lost or greatly reduced the paraconid. Several primates have molarized premolars (see fig. 6.9), for example, P4 in *Galago, Arctocebus, Eulemur,* or *Hapalemur.* Folivorous primates like the indriids (fig. 6.15), *Alouatta,* or *Brachyteles* have enlarged upper molars with a hypocone and their molar cusps are cresty, often with buccal styles (i.e., para-, meso-, and metastyles). In contrast, the sclerocarp feeders like the pitheciines often possess small premolars and molars but prominent anterior dentitions with enlarged or procumbent upper and lower incisors often with stout upper canines. Frugivores are similar

in promoting their anterior dentition but less extreme in molar reduction relative to pitheciines. Gum feeders like *Cebuella* and *Callithrix* possess tall incisors like their canines, whereas *Euoticus, Phaner,* and galagos use toothcombs.

Anthropoid teeth are more similar anatomically than those of the strepsirhines. All anthropoids have spatula-shaped central and lateral incisors and tall canines (figs. 6.16–19). Canine size is often sexually dimorphic, with males possessing longer and larger canines (see fig. 6.8). Platyrrhines in general possess hypocones, like other anthropoids, but have a unique strong lower molar crest, the protocristid, which connects the protoconid with the metaconid and separates

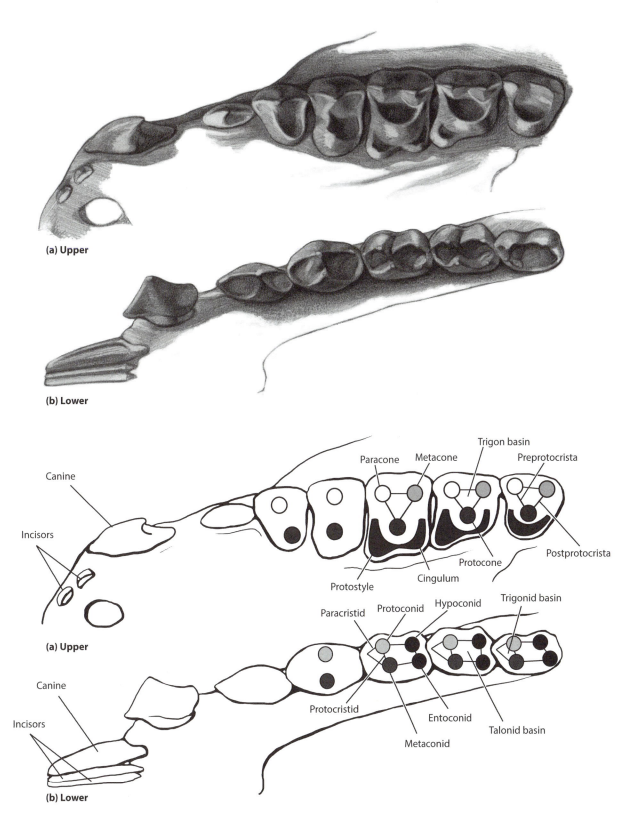

Figure 6.14 Upper and lower dentition of *Eulemur fulvus*.

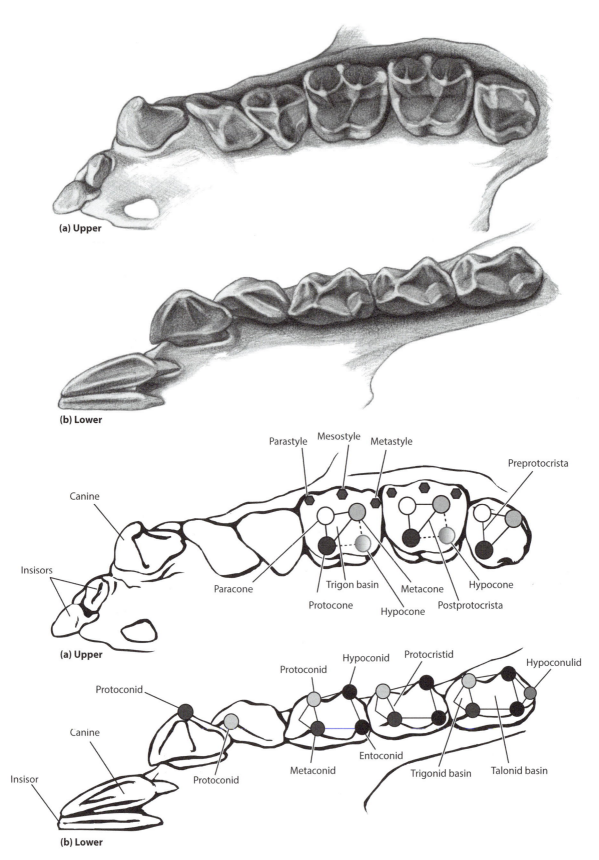

Figure 6.15 Upper and lower dentition of *Propithecus verreauxi*.

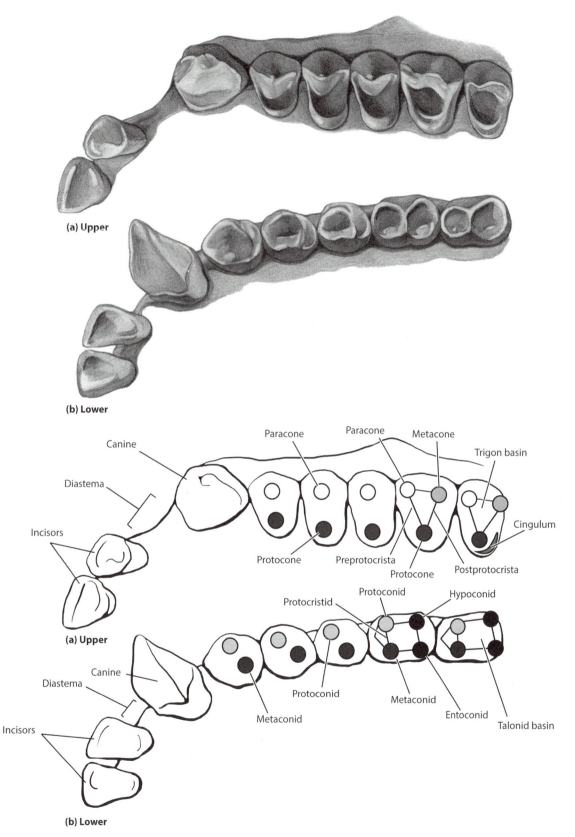

Figure 6.16 Upper and lower dentition of *Saguinus oedipus*.

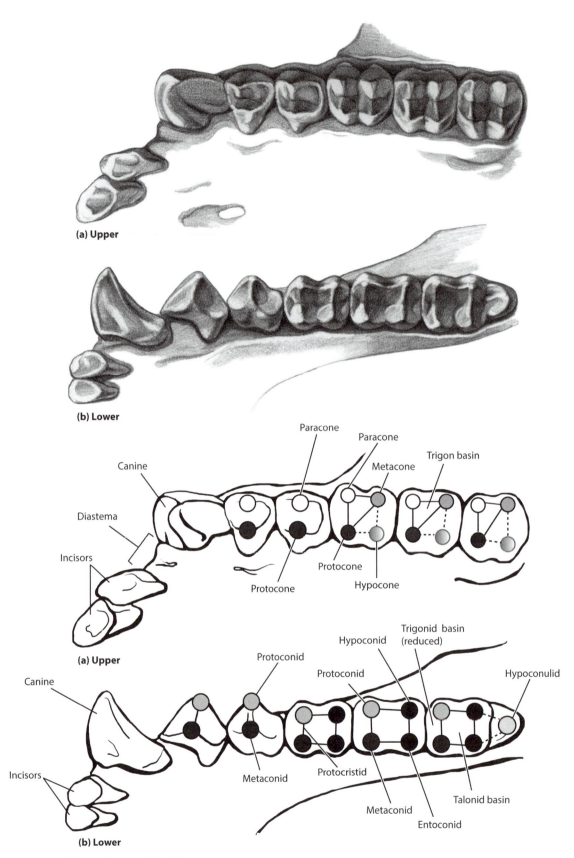

Figure 6.17 Upper and lower dentition of *Colobus guererza*.

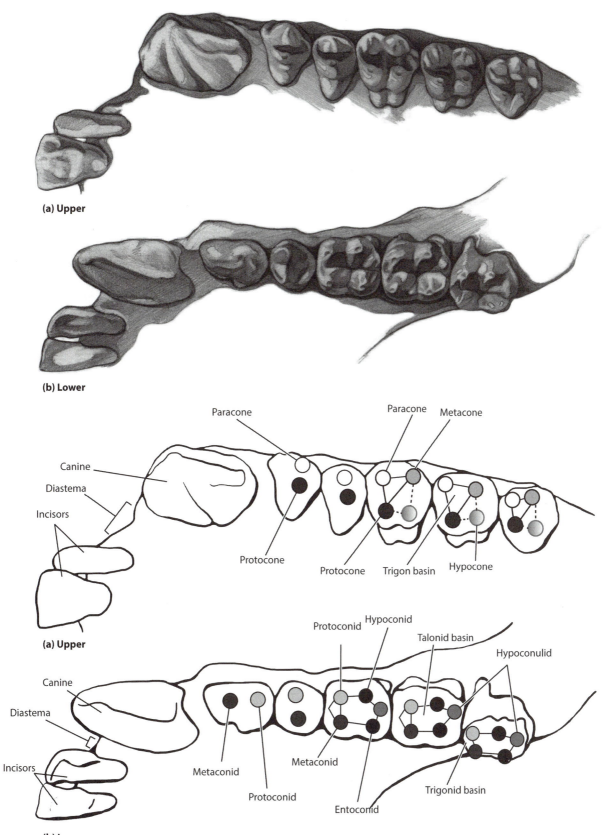

(a) Upper

(b) Lower

Paracone

Paracone   Metacone

Canine

Diastema

Incisors

Protocone

Protocone   Trigon basin   Hypocone

(a) Upper

Protoconid   Hypoconid

Talonid basin

Hypoconulid

Canine

Diastema

Incisors

Metaconid

Metaconid

Protoconid

Entoconid

Trigonid basin

(b) Lower

Figure 6.18 Upper and lower dentition of *Pongo pygmaeus*.

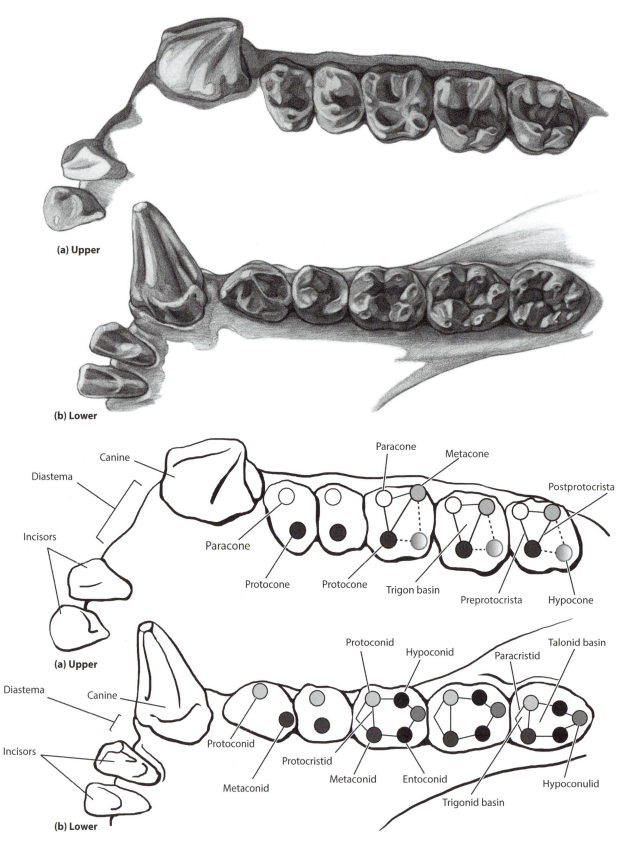

Figure 6.19 Upper and lower dentition of *Gorilla gorilla*.

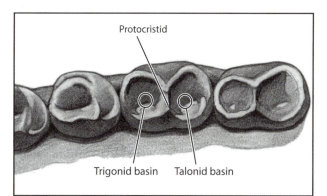

(a) *Saguinus* (lower molars + premolars)

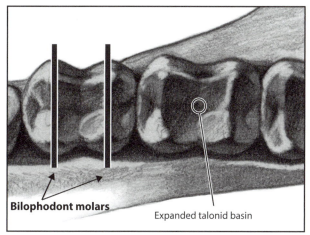

(b) *Colobus* (lower molars)

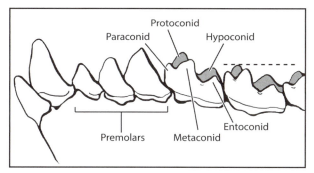

(c) *Tarsius* (lingual view)

Figure 6.20 Unique molar features in primates: (a) the protocristid crest that connects the protoconid with the metaconid in platyrrhines; (b) bilophodont molars of Old World monkeys; and (c) a tall trigonid relative to the talonid basin in tarsiers.

the trigonid from the talonid basin (fig. 6.16). The trigonid basin is reduced in platyrrhines, reduced even more in apes, and lost in Old World monkeys. All Old World monkeys possess an unusual cusp arrangement on their molars called bilophodonty (fig. 6.20), and their long, dagger-like canines normally possess

a vertical groove (the mesial sulcus). Surprisingly, callitrichids, which are anthropoids, possess tribosphenic molars (without hypocones), as do many strepsirhines.

THE AYE-AYE

Aye-ayes have greatly reduced their dentition and they possess enlarged, rodent-like upper and lower incisors (figs. 5.11 and 6.4). *Daubentonia* has tiny molars and upper P4. Match these dental characters to their robust mandible and compact skull and you can see that the diet and the processing of food items by aye-ayes must be quite different from that of other lemurs. In fact, aye-ayes consume structurally defended foods, usually larvae hidden within the bark of tree branches. Aye-ayes use their large ears to locate larvae and their robust incisors to make openings in the bark where they can insert their slender and elongated third manual digit to spear beetle larvae. Because they consume soft-bodied food items, aye-ayes have little use for molar crushing and grinding, thus their reduced molar sizes.

DIET: FRUITS, LEAVES, AND INSECTS

Primates generally eat three major food types: insects (insectivores), fruits (frugivores), and leaves (folivores) (table 6.3). Usually, two of these items predominate in a species-specific diet. For example, large primates like chimpanzees generally prefer fruits and then leaves, while small primates prefer insects (i.e., arthropods). Fruits are high-quality carbohydrates—energy stores with calories that can be burned quickly. Leaves have a higher protein content, the building blocks of cells, but they are digested slowly. Insects are high in both carbohydrates and proteins. Both carbohydrate and protein sources need to be balanced for an adequate primate diet. Primates can only digest leaves with the aid of microorganisms within their guts since mammals cannot break down cellulose, the cell wall of plant cells. A few primates specialize on lichens, seeds, gums, or grass (specialized folivores), and a few even eat live prey occasionally (carnivory).

Small and large primates use food resources differently due to an interesting interaction between body size and metabolism. Metabolism, or basal metabolic

Table 6.2 Specialized Dental Characteristics across Selected Primates

Primates	Dental Features
Cheirogalidae	Tribosphenic molars, toothcomb, protocristid connects protoconid and metaconid, lower molar paraconids are absent
Phaner	Large upper central incisor and canine, caniniform lower P2, hypocones
Lemuridae	Tribosphenic molars, toothcomb, lower molar paraconids are absent
Eulemur	Protostyles on upper molars
Hapalemur	Molarized upper P4 with mesostyles, small protostyles and hypocones
Lemur	Reduced upper incisors
Indriidae	Enlarged upper molars, hypocone, cresty cusps and buccal styles, lower molar paraconids are absent, upper lateral incisors are small, toothcomb
Avahi	Small upper lateral incisors
Lepilemuridae	Large hypoconid, large talonid basin, loss of upper incisors, lower molar paraconids are absent, toothcomb
Daubentoniidae	Greatly reduced dentition, rodent-like incisors
Lorisidae	Hypocones are common, lower molar paraconids are absent, toothcomb, reduced upper M3s
Arctocebus	Hypocone with a molarized upper P4
Galagidae	Hypocones are common, sharp molar cusps, molarized upper P4, lower molar paraconids are absent, toothcomb
Tarsiidae	Paraconids, tribosphenic molars with steep pointed cusps, loss of a lower incisor, pointed incisors
Platyrrhini	Hypocones except for the callitrichids, a well-developed protocristid, reduced trigonic basin
Callitrichidae	Tribosphenic molars, loss of upper and lower M3s (except *Callimico*), reduced M2s relative to M1s, mesostyles are common
Callithrix	Tall lower incisors
Cebuella	Tall lower incisors
Callimico	Upper molar styles, small M3s, especially the upper M3
Pitheciinae	Premolar and molar reduced relative to enlarged anterior teeth
Cebus	Small M3s
Alouatta	Enlarged upper molars with styles and cresting for folivory, small upper M3
Brachyteles	Enlarged upper molars with styles and cresting for folivory, small upper M3
Cercopithecidae	Bilophodont upper and lower molars, loss of trigonid basin, dagger-like canines with a mesial sulcus, elongated or sectorial lower P3
Colobinae	Taller molar cusps and smaller central incisors, hypoconulids on lower M3s
Hylobatidae	Broad central incisors, Y-5 lower molar cusp pattern, reduced trigonid basin
Pongidae	Broad central incisors, Y-5 lower molar cusp pattern, reduced trigonid basin
Pongo	Procumbent upper incisors, molar wrinkling, thicker molar enamel
Pan	Small molars
Gorilla	Large premolars and molars with high cusps

rate, is much higher for small primates, and this rate decreases as primates, or mammals, increase in size. Thus, small primates need to feed often to keep their metabolism functioning adequately, while large primates eat a lot of bulk but generally consume lower-quality food items. In short, larger primates eat more and move very little. Consider the hay that horses eat versus the sugar water that keeps hummingbirds in flight. Of the three food types that primates feed on, insects and leaves are quite abundant while fruit is scattered among the trees in a forest (a patchy food item). Although insects are reasonably abundant, they are generally solitary food items and are also spread throughout the spatial environment, representing a patchy food item as well. In contrast, leaves are more uniform in their distribution—although they also have their own chemical defenses against predators. Thus, although insects are quite nutritious, "tiny

candy bars," if you will, they are difficult to catch in abundance; while leaves are easy to obtain, they are generally poor in carbohydrates. Leaves also have the distinct disadvantage of taking a lot of time to digest and they represent a low-quality food item. Calorie return is slow for folivores. This means that only small primates (< 500 g) can find enough insects to be truly insectivorous in their dietary choices since the rate of insect capture does not increase with increased body size. As body size increases for insect-eating primates, these primates cannot catch enough insects to survive so they are "forced" to switch to gums or to fruit (both are carbohydrate sources) in combination with insects to stay alive.

In contrast, as folivores decrease in size, they face a different energetic problem. Leaves digest so slowly that as primates get smaller (< 700 g), small folivores cannot retrieve enough calories fast enough from

Table 6.3 Primate Diets, in %

Taxa	Fruit	Leaves	Insects	Other
Arctocebus calabarensis	14		85	
Perodicticus potto	65		10	21 (gums)
Loris tardigradus			100	
Nycticebus coucang	23		3	75 (gums and nectar)
Galago demidovii	19		70	10 (gums)
Galago moholi			52	48 (gums)
Otolemur garnettii	50		50	
Euoticus elegantulus	5		20	75 (gums)
Mirza coquereli	30		50	20 (gums)
Eulemur fulvus	25	71		4 (flowers)
Lemur catta	34	44		23 (flowers and herbs)
Lepilemur mustelinus		51		48 (flowers)
Propithecus verreauxi	46	33		
Tarsius bancanus			100	
Cebuella pygmaea			33	67 (gums)
Callithrix jacchus	18		4	76 (gums)
Callimico goeldii	29		33	29 (fungus)
Saguinus geoffroyi	60	10	30	
Cebus capucinus	65		20	
Callicebus torquatus	67		14	
Pithecia pithecia	31	10	4	53 (seeds)
Aotus azarai	45	41		14 (nectar and flowers)
Alouatta palliata	29	49		23 (flowers)
Ateles geoffroyi	80	20		
Cercopithecus ascanius	61	7	25	5 (gums and flowers)
Cercopithecus diana	36	8	25	30 (seeds and flowers)
Macaca fascicularis	67	17	4	9 (flowers and herbs)
Macaca mulatta	9	84		6 (flowers and herbs)
Papio anubis	12		2	80 (herbs)
Pilocolobus badius	36	47		9 (flowers)
Colobus guereza	5	86		1 (flowers)
Colobus satanas		39		53 (seeds)
Presbytis entellus	45	48		
Rhinopithecus bieti		6		86 (lichens)
Hylobates lar	66	24	9	1 (flowers)
Hylobates syndactylus	45	44	8	4 (flowers)
Pongo pygmaeus	54	29	1	
Pan troglodytes	68	28	4	
Gorilla gorilla	2	86		5 (herbs and flowers)

Modified from Richard, 1985, and Campbell et al., 2007.

a leafy diet to maintain their higher metabolic rate. Thus, folivores tend to be larger primates with large guts (or storage chambers) for digestion (e.g., the big belly of gorillas).

The digestive tract of primates is related to their specific dietary preferences. Insect-eating primates have simple and short guts (i.e., a stomach, small and large intestines, and a caecum). Some primates have large stomachs (foregut fermenters); others have expanded the caecum (a pouch at the junction of the small and the large intestines) and are known as hindgut fermenters (fig. 6.21). Colobine monkeys are foregut fermenters, while indriids, *Lepilemur*, and many of the gum-feeding galagos are hindgut fermenters with enlarged caecums. Fruit eaters tend to have a long small intestine.

ODDITIES

Anatomical features related to diet include the cheek pouches of cercopithecine Old World monkeys. These monkeys fill their cheek pouches quickly with food items until they are bulging. They then move to later chew and digest these items away from their gathering point. Old World baboons are dietary specialists for eating grasses (herbivores), while the New World pitheciines are seed specialists. Sclerocarp feeders, or seed eaters like the pitheciines, show outwardly flared and robust canines with procumbent lower and upper incisors, as noted above. This modified anterior dentition gives sclerocarp feeders an unusual facial morphology. Seed eaters also possess generally smaller and flattened molars with thicker enamel relative to other

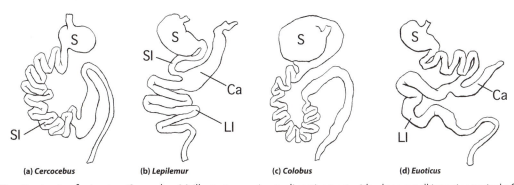

(a) *Cercocebus*   (b) *Lepilemur*   (c) *Colobus*   (d) *Euoticus*

Figure 6.21 Digestive tracts of primates. *Cercocebus* (a) illustrates a primate digestive tract with a long small intestine typical of a frugivorous primate. *Lepilemur* (b) shows a long and large caecum typical of a hindgut fermenter. *Colobus* (c) displays a large stomach and is a foregut fermenter. *Euoticus* (d) is gumnivorous and its digestive tract also shows an enlarged caecum. S = stomach; SI = small intestine; Ca = caecum; and LI = large intestine. Modified from Fleagle, 1999.

primate molars. Perhaps the most unusual primate diets are the cyanide-laced bamboo diet of *Hapalemur*, the lichen diet of *Rhinopithecus*, and the fungi diet of *Callimico*.

Selected References

Ankel-Simons, F. 2000. Primate Anatomy—An Introduction. 2nd edition. Academic Press, New York.

Burrows, A.M., and L.T. Nash. 2010. The Evolution of Exudativory in Primates. Springer, New York.

Campbell, C.J., A. Fuentes, K.C. Mackinnon, M. Panger, and S.K. Bearder (eds.). 2007. Primates in Perspective. Oxford University Press, Oxford, UK.

Cartmill, M.C. 1974. *Daubentonia, Dactylopsila*, woodpeckers and klinorhynchy; pp. 655–672 *in* R.D. Martin, G.A. Doyle, and A.C. Walker (eds.), Prosimian Biology. Duckworth Publishing, London.

Charles-Dominique, P. 1977. Ecology and Behaviour of Nocturnal Primates. Columbia University Press, New York.

Chivers, D.J., B.A. Wood, and A. Bilsborough (eds.). 1984. Food Acquisition and Processing in Primates. Plenum Press, New York.

Fleagle, J.G. 1999. Primate Adaptation and Evolution. Academic Press, New York.

Glander, K.E., P.C. Wright, P.C., D.S. Seigler, and B. Randriansolo. 1989. Consumption of cyanogenic bamboo by a newly discovered species of bamboo lemur. American Journal of Primatology 19:199–224.

Groves, C. 2001. Primate Taxonomy. Smithsonian Institution Press, Washington, D.C.

Hershkovitz, P. 1977. Living New World Monkeys (Platyrrhini). University of Chicago Press, Chicago.

Hillson, S. 1996. Dental Anthropology. Cambridge University Press, Cambridge, UK.

Hladik, C.M., P. Charles-Dominique, and J.J. Peter. 1980. Feeding strategies of five nocturnal prosimians in the dry forest of the West Coast of Madagascar; pp. 41–74 *in* P. Charles-Dominique, H.M. Cooper, A. Hladik, C.M. Hladik, E. Pages, G.E. Pariente, A. Petter-Rousseaux, and A. Schilling (eds.), Nocturnal Malagasy Primates. Academic Press, New York.

Hylander, W.L. 1975. Incisor size and diet in anthropoids with special reference to the Cercopithecidae. Science 189:1095–1098.

Jablonski, N.G., and R.H. Crompton. 1994. Feeding behavior, mastication, and tooth wear in the Western Tarsier (*Tarsius bancanus*). International Journal of Primatology 15:29–59.

James, W.W. 1960. The Jaws and Teeth of Primates. Pitman Medical Publishing Company, London.

Kay, R.F. 1975. The functional adaptations of primate molar teeth. American Journal of Physical Anthropology 43:195–216.

———. 1984. On the use of anatomical features to infer foraging behavior in extinct primates; pp. 21–53 *in* J. Cant and P. Rodman (eds.), Adaptations for Foraging in Nonhuman Primates. Columbia University Press, New York.

Kay, R.F., and H.H. Covert. 1984. Anatomy and behavior of extinct primates; pp. 467–508 *in* D.J. Chivers, B.A. Wood, and A. Bilsborough (eds.), Food Acquisition and Processing in Primates. Plenum Press, New York.

Kinzey, W.G. 1992. Dietary and dental adaptations in the Pithecinae. American Journal of Physical Anthropology 88:499–514.

Kirkpatrick, R.C., Y.C. Long, T. Zhong, and L. Xiao. 1998. Social organization and range use in the Yunnan snub-nosed langur *Rhinopithecus bieti*. International Journal of Primatology 70:117–120.

Maier, W. 1984. Functional morphology of the dentition of the Tarsiidae; pp. 45–58 *in* C. Niemitz (ed.), Biology of Tarsiers. Gustav Fischer Verlag, Stuttgart, Germany.

———. 1984. Tooth morphology and dietary specialization; pp. 303–330 *in* D.J. Chivers, B.A. Wood, and A. Bilsborough (eds.), Food Acquisition and Processing in Primates. Plenum Press, New York.

Martin, R.D. 1990. Primate Origins and Evolution—A Phylogenetic Reconstruction. Princeton University Press, Princeton, N.J.

Plavcan, J.M. 1993. Canine size and shape in male anthropoid primates. American Journal of Physical Anthropology 92:201–216.

Plavcan, J.M., and C.P. Van Schaik 1994. Canine dimorphism. Evolutionary Anthropology 2:208–214.

Porter, L.M. 2007. The Behavioral Ecology of Callimicos and Tamarins in Northwestern Bolivia. Pearson Prentice Hall, Upper Saddle River, N.J.

Richard, A.F. 1985. Primates in Nature. W.H. Freeman and Company, New York.

Rodman, P.S., and J.H. Cant (eds.). 1984. Adaptations for Foraging in Nonhuman Primates. Columbia University Press, New York.

Rosenberger, A.L., and W.G. Kinzey. 1976. Functional patterns of molar occlusion in platyrrhine primates. American Journal of Physical Anthropology 45:281–298.

Rosenberger, A.L., and E. Strasser. 1985. Toothcomb origins: support for the grooming hypothesis. Primates 26:73–84.

Seligsohn, D. 1977. Analysis of species-specific molar adaptation in strepsirhine primates. Contributions to Primatology 11:1–116.

Smuts, B.B., D.L. Cheney, R.M. Seyfarth, R.W. Wrangham, and T.T. Struhsaker (eds.). 1986. Primate Societies. University of Chicago Press, Chicago.

Sterling, E.J., E.S. Dierenfeld, C.G. Ashbourne, and A.T.C. Feistner. 1994. Dietary intake, food composition and nutrient intake in wild and captive populations of *Daubentonia madagascariensis*. Folia Primatologica 62:115–124.

Swindler, D.R. 2002. Primate Dentition—An Introduction to the Teeth of Non-Human Primates. Cambridge University Press, Cambridge, UK.

Szalay, F.S., and E. Delson. 1979. Evolutionary History of the Primates. Academic Press, New York.

Teaford, M.F., M.M. Smith, and M.W.J. Ferguson (eds.). 2000. Development, Function, and Evolution of Teeth. Cambridge University Press, Cambridge, UK.

Unger, P.S. 1994. Patterns of ingestive behavior and anterior tooth use differences in sympatric anthropoid primates. American Journal of Physical Anthropology 95:197–219.

———. 2010. Mammal Teeth. Johns Hopkins University Press, Baltimore.

Whiten, A., and M. Widdowson (eds.). 1992. Foraging Strategies and Natural Diet of Monkeys, Apes and Humans. Oxford Science Publications, Oxford, UK.

7 | Backs

TUBES

Most primates, like quadrupedal mammals, have a thorax that is shaped like a tube or a cylinder (fig. 7.1). The forelimbs are set below and close to the midline of the body, while the shoulder blades (scapulae) lie along the sides of the thoracic cylinder. Heads and tails span each end with the viscera hanging down from the internal dorsal wall. This is a long body type that is held together by the vertebral column. The number of vertebrae in primates varies, allowing for vertebral length changes and back mobility across groups. In terms of vertebrae (figs. 7.1–7.3), primates always have 7 cervical vertebrae in the neck. Numbers of the other vertebrae vary more widely: thoracic vertebrae, 11–14; lumbar vertebrae, 5–9; and sacral vertebrae, 3–7. Almost all primates have tails and therefore possess caudal vertebrae; the living apes, including humans, are exceptions. Lorises, *Indri, Cacajao*, and pig-tailed macaques are also known for their reduced tail lengths. In contrast, spider monkeys and their close kin are renowned for their long, muscular, and prehensile grasping tails, an adaptation that is not common among primates.

VERTEBRAL COLUMN

The vertebral column (fig. 7.2) and its corresponding vertebrae (fig. 7.3) act as the central anchor or beam for all of the other body parts. In a primate vertebra the vertebral centrum is the large joint surface that absorbs and transmits the weight-bearing forces along the entire vertebral column of the primate body (fig. 7.4). Intervertebral disks are interspaced between the vertebrae; they are made of fibrocartilage and represent a softer, more flexible padding between the hard bony vertebral bodies. These disks act as shock absorbers. The vertebral joint surfaces and their bony processes reflect the horizontal back posture, called pronogrady, which is exhibited by most primates. These bony processes extend away from vertebrae and serve to anchor vertebral ligaments and the long back muscles arrayed along the entire vertebral column (i.e., the flexible beam part of the back).

Along the entire sequence of vertebrae lies a complex series of ligaments that hold these axial elements together (fig. 7.5). The anterior and posterior ligaments extend along the length of the vertebral bodies and disks and they guard against excessive movements of this semi-flexible column. The ligamentum flavum attaches each spinous process to a transverse process. The supraspinous ligaments line up along the top of the spinous process, while the interspinous ligaments connect each spinous process sagittally. Ligaments bind the vertebrae together, but they also allow a certain amount of flexibility in the backs of primates, particularly those with pronograde body postures.

NECKS, BACKS, AND TAILS

In the primate neck, there are only seven cervical vertebrae, and the first two possess a different anatomical

Scientists need reference directions to facilitate communication about the location of one body part relative to another. We have illustrated common orientation terms for a primate body, *Eulemur* in this case.

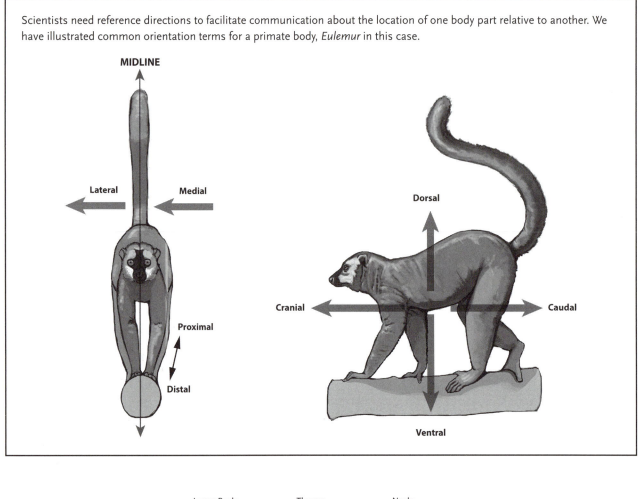

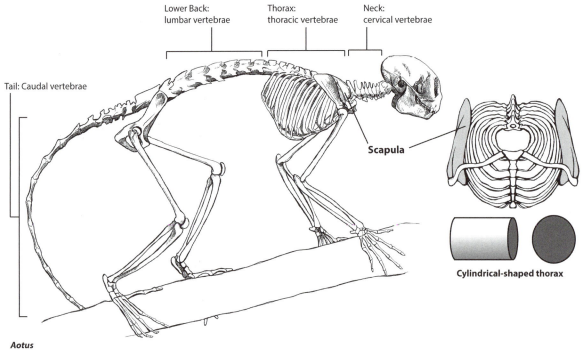

Figure 7.1 A primate skeletal body with a cylindrical-shaped thorax (pronogrady). Note the position of the scapula along the sides of the thorax.

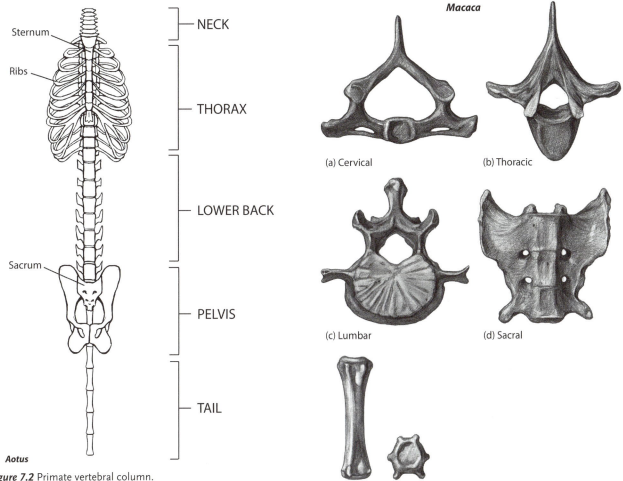

Figure 7.2 Primate vertebral column.

Aotus

Labels in Figure 7.2: Sternum, Ribs, Sacrum; NECK, THORAX, LOWER BACK, PELVIS, TAIL

Macaca

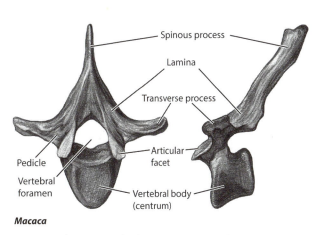

(a) Cervical   (b) Thoracic

(c) Lumbar   (d) Sacral

(e) Caudal (dorsal and distal views)

Figure 7.3 Primate vertebrae (*Macaca*): (a) cervical; (b) thoracic; (c) lumbar; (d) sacral; and (e) caudal.

form from the other five (fig. 7.6). The atlas, cervical vertebra number one, is shaped like an oblong ring; this vertebra articulates with the occipital condyles at the base of the head, allowing head flexion and extension. As a ring it has lost its centrum. The second of the cervical vertebrae, the axis, has added the centrum from C1 into a bony vertical process called the dens, and C1 rotates around this vertical pole. The dens on the axis cervical vertebra (fig. 7.6) is oriented dorsally in quadrupedal primates, is only slightly bent in African apes, and appears straight in the upward vertebral column of humans. According to Friderun Ankel-Simons of Duke University, the position of the dens reflects the position of the head relative to the back. In tarsiers, the articular facets of the cervical vertebrae are positioned perpendicular to the vertebral column in such a way that tarsiers can turn their heads around 180 degrees to face backward relative to their chests.

Labels in Figure 7.4: Spinous process, Lamina, Transverse process, Articular facet, Pedicle, Vertebral foramen, Vertebral body (centrum)

Macaca

Figure 7.4 Thoracic vertebral structures (*Macaca*).

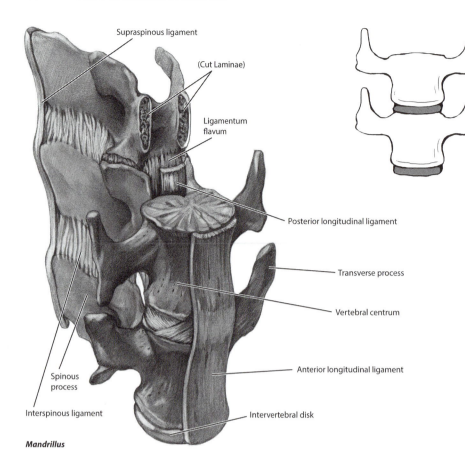

Mandrillus

Figure 7.5 Vertebral ligaments. The upper right image illustrates the intervertebral disks (shaded) between vertebrae.

Another notable neck modification in primates includes that in male gorillas and orangutans, which have especially elongated cervical spinous processes (fig. 7.7) for the large neck muscles that hold up their massive heads. Lorises also have elongated cervical spinous processes (fig. 7.7) but possess small heads relative to those of the great apes. In the potto, the last two cervical and the first two thoracic vertebrae have elongated spinous processes into dorsal spines that are associated with a cornified epithelium that functions as a sensory zone.

As we move caudally downward from the neck, the primate thorax holds all of the rib-bearing vertebrae. The ribs curve gently to form the long, cylindrical back of primates, encapsulating the heart and lungs (see fig. 7.2). On the ventral surface of the thorax, the sternum (fig. 7.8; see also fig. 7.2) is narrow with several bony elements, typically six. Along the side of the thorax there are three types of ribs: true, false, and floating. True ribs articulate with the sternum via cartilage at the sternal notches. False ribs attach to separate strands of cartilage that merge before

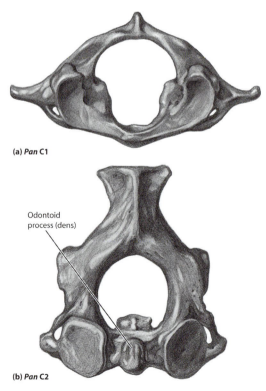

(a) *Pan* C1

(b) *Pan* C2

Figure 7.6 Atlas (*top*) and axis (*below*), the first two cervical vertebrae.

contacting the ventral region of the sternum. Floating ribs do not articulate with the sternum at all and are located on the last of the thoracic vertebrae. In great apes and humans, the diaphragm lies at the level of the last thoracic vertebra in contrast to other primates, where rib-bearing vertebrae extend caudally to the diaphragm.

The lower back is the region where the lumbar vertebrae occur. They are fairly numerous, long, and flexible across pronograde primates and can be defined as non-rib-bearing vertebrae before the sacrum (figs. 7.1–7.3). Lumbar vertebrae are quite variable in terms of numbers across primates from a low of four in the great apes to nine in *Indri* and *Lepilemur*. Some primates have expanded lumbar spinous processes (e.g., *Ateles*), while many have shape or angle of orientation changes for these dorsally oriented spinous processes. Lumbar vertebrae may possess accessory processes as attachment sites for tail and back muscles. These accessory processes are located on the dorsal lamina, where they can be relatively large and pointed. The accessory processes tend to diminish in size as they progress caudally.

Among strepsirhines, the shortest lumbar regions belong to galagos, *Nycticebus* and *Perodicticus*, and indriids, while the longest sections are found among lemurs, cheirogaleids, and *Lepilemur*. The long lumbar taxa have elongated vertebral bodies, suggesting a greater range of motion in the lower back for these primates. This group also shares relatively long spinous processes that serve as attachment sites for extensive lower back extensor musculature. The vertical clinging and leaping primates, the indriids and galagos, both possess short lumbar regions but they differ in spinous process size and orientation, being short and dorsally oriented among indriids.

The sacrum is composed of several, often fused,

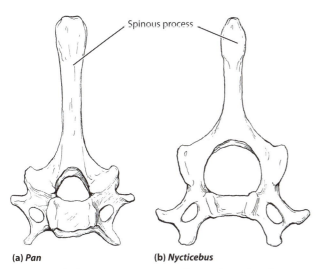

(a) Pan   **(b) Nycticebus**

Figure 7.7 The long spinous process of cervical vertebrae in apes and lorises.

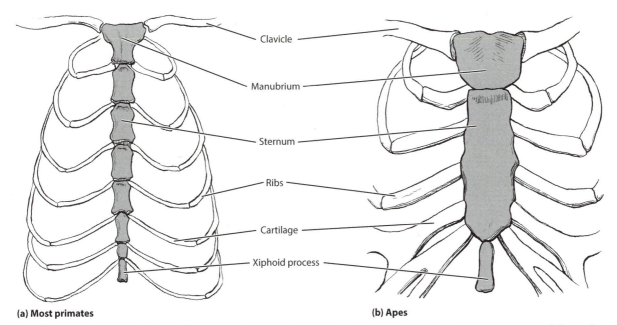

(a) Most primates   **(b) Apes**

Figure 7.8 Sternum. There are structural differences between most primates (a) and apes (b). Note the more numerous and thinner bony segments among most primates.

vertebral elements, making this section of the vertebral column the largest (figs. 7.2 and 7.3). Sacral elements are usually 3 in number but lorises have 7 and apes 5 or 6. The sacrum's most important function is to connect the pelvis and caudal vertebrae to the vertebral column. The neural arches and spines are commonly fused on the dorsum of the sacrum. The size of the sacral foramen is related to the length and functional capacity of the tail.

Tails come in several varieties among the living primates (fig. 7.9). The longest tails tend to be found among the frequent leaping primates as well as the prehensile-tailed monkeys (atelines or the spider monkey group). Some primates, like the apes, have lost their tails, while lorises, *Indri, Cacajao*, mandrills, drills, and some macaques have greatly reduced tails. The prehensile-tailed atelines all possess a frictional pad with dermatoglyphics along the ventral surface of their long and muscular tail to wrap around branches. This prehensile tail acts as a fifth grasping limb for these suspensory primates. In the prehensile-tailed atelines, the sacral openings are large as are the neural arches, thus the nerve connections and muscular attachments are better developed among these taxa. *Cebus* also possesses a prehensile tail but without a frictional surface.

Most leaping primates have long tails that help in the aerodynamics of balance during leaping. The tail plays an active role as a steering mechanism during leaping and can help mid-flight body rotations prior to landing. In contrast, the vertical clinging and leaping tarsiers have thinly haired tails with distal tufts that appear more rat-like in appearance. Tarsiers can even use their tails to support their body during vertical clinging. Some primate tails, like that of the ring-tailed lemur, are banded with black and white rings or are exceedingly bushy as in the aye-aye (fig. 7.9).

Tail, or caudal, vertebrae are typically small and variable in number. Caudal vertebrae number in the twenties for most primates but only 9 in lorises and 3 in apes. These vertebrae typically get smaller as they progress caudally away from the sacrum. The caudal vertebrae are generally uniform in morphology among most primates (see fig. 7.3). Prehensile-tailed monkeys have the longest caudal element in a more distal position relative to other long-tailed primates. Chevron bones, or the small V-shaped bones that attach to the ventral surface of the more proximal caudal vertebrae, allow the caudal artery to pass through these small bony structures to supply the soft tissues of the tail.

BOARDS

The thorax of living apes, including humans, is shaped differently from that of all other pronograde primates

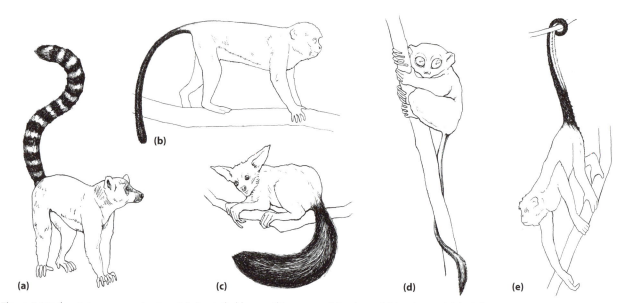

Figure 7.9 Tail variety among primates: (a) ring-tailed lemur; (b) guenon; (c) galago; (d) tarsier; and (e) atelines.

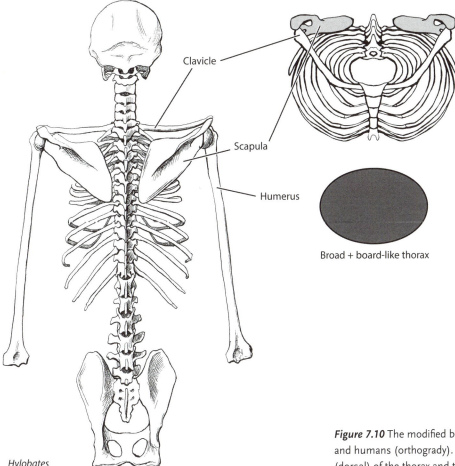

Clavicle

Scapula

Humerus

Broad + board-like thorax

Hylobates

Figure 7.10 The modified broad, or board-like, thorax of apes and humans (orthogrady). The scapulae lie on the back surface (dorsal) of the thorax and the clavicles are long.

(fig. 7.10). Ape backs are flattened and compressed dorsoventrally while being broad mediolaterally, more like a flattened board than a cylindrical tube. The clavicles are long and broaden toward the shoulder. Scapular placement is on the flat of the back rather than along the sides of a tube. This position forces the arms to come off the far lateral sides of the thorax. Pushing the shoulders away from the midline of the body and toward the sides of the body is an odd adaptation relative to other primates or mammals. This location adds great mobility at the shoulder joint in apes and humans, thereby giving these specialized primates arms capable of circling above the head (circumduction).

The back of an ape is more erect relative to that of other primates. Being erect requires the heart and lungs to be attached to the dorsal body wall, rather than hanging free and downward within the thorax cavity as in pronogrady. Likewise, the large intestine

has mesenteric attachments to keep it from falling out of the abdominal cavity. The soft tissues in ape and human bodies need to be anchored due to this body orientation, which is called orthogrady.

Orthograde backs require the ribs to be more curved (fig. 7.11), either flaring outward at the bottom like a funnel or being tucked inward like those of gibbons or humans. The sternum is much shorter and broader in apes relative to the long and narrow condition found in other primates (see fig. 7.8).

In orthograde ape bodies, the lower back is stiff and the lumbar vertebrae are shorter relative to the long and more flexible backs among other primates (fig. 7.12). The lumbar spinous process is short, flatter, and oriented more perpendicularly in apes relative to that of pronograde primates. Likewise, the transverse processes are positioned higher in apes relative to those in pronograde primates. In apes, the transverse vertebral process comes off just above the vertebral body at the vertebral body and pedicle junction or above at the pedicle-lamina border, being laterally or dorsally oriented. These lateral- or dorsal-oriented processes are positioned to maintain back muscle function relative to the wide iliac blades and ribs in apes. In other primates, the transverse process comes off lower along the side of the vertebral body and this process is oriented ventrally (fig. 7.13). The ventral displacement of this transverse process functions to create deep erector spinae musculature, muscles that are enlarged in pronograde primates relative to apes. Atelines are similar to gibbons in the orientation and exit point of the transverse vertebral process.

In apes, the lumbar centrum is broader and round relative to the more kidney-bean joint surface shape found among other primates (fig. 7.13). These larger vertebral joints are aligned along the erect back of apes and they imply greater back forces given their extra bone development. The lumbars are also shorter in apes as is the entire hominoid lumbar region relative to that of other primates.

Ateles and *Brachyteles*, New World brachiators, also share lumbar adaptations that resist bending and both possess the relatively shortest lumbar vertebral bodies and lumbar regions of all atelines. These New World brachiators are similar to apes in spinous process

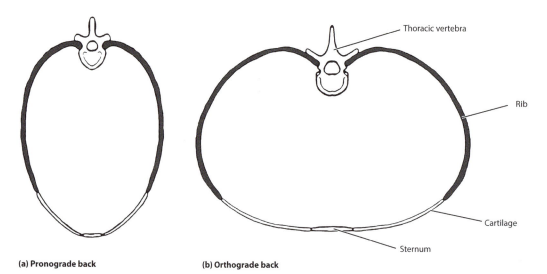

(a) Pronograde back   **(b) Orthograde back**

Figure 7.11 Rib curvature among primates: (a) pronograde and (b) orthograde back plans. Ribs are shaded.

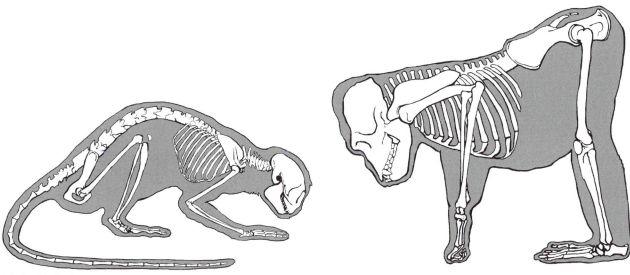

(a) Flexible back   **(b) Stiff lower back**

Figure 7.12 Back flexion among primates: (a) the flexible and curved back of most primates; and (b) the stiff lower back of apes.

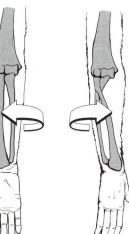

Adduction Abduction

Dorsiflexion Plantarflexion

Supination Pronation

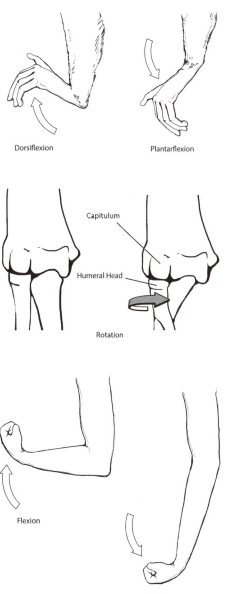

Capitulum

Humeral Head

Rotation

Flexion

Eversion Inversion

Extension

Primate bodies, including their limbs, move in many directions, and the study of these motions requires a terminology that allows anatomists to understand the direction of one movement relative to another. The common terms for body motions are listed below.

Abduction: a movement away from the midline or sagittal plane.

Adduction: a movement toward the midline, the opposite of abduction.

Dorsiflexion: flexion of the foot toward the lower leg.

Plantarflexion: flexion of the foot downward, toward the ground, the opposite of dorsiflexion.

Supination: a rotary motion that lays the hand on its back side, palm up.

Pronation: a rotary motion that positions the palm of the hand downward, the opposite of supination.

Rotation: a turning of a body part (such as the radius) on an axis.

Inversion: a rotary motion of the foot that orients the sole medially.

Eversion: a rotary motion of the foot that orients the sole outward, the opposite of inversion.

Flexion: a bending motion that brings body parts closer together.

Extension: a straightening motion that is the opposite of flexion.

Two back types, pronograde and orthograde, occur across living primates. One of the distinctive morphological features for these two back types occurs on the side of lumbar vertebrae and involves the location of the transverse process. In pronograde, or monkey-like, horizontally adapted backs, the transverse process diverges from the lumbar vertebral centrum, a low position, in contrast to orthograde, or ape-like, vertically adapted backs where the transverse process has shifted proximally and diverges from the dorsal lamina of a lumbar vertebra. Illustrated below are two lumbar vertebrae from two early Miocene fossil "apes," *Proconsul africanus* from Kenya (a) and *Morotopithecus bishopi* from Uganda (b). Note the position of the transverse process in both lumbar vertebrae for these two fossil primates. In *Proconsul*, the location of the transverse process is similar to that of pronograde monkeys like *Papio*. The transverse process exits much higher for *Morotopithecus*, looking anatomically more similar to the morphological condition found in *Pan* (compare with fig. 7.13), suggesting that *Morotopithecus* has an orthograde or an ape-like vertically oriented back relative to the monkey-like back of *Proconsul*.

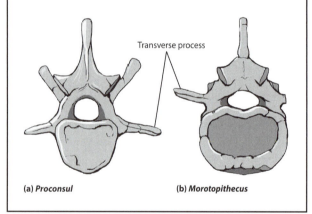

(a) *Proconsul*   (b) *Morotopithecus*

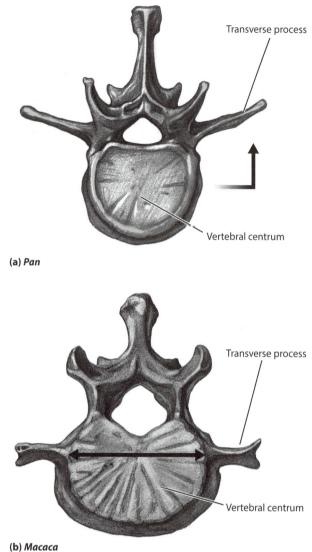

(a) *Pan*

(b) *Macaca*

Figure 7.13 Lumbar vertebral differences between monkeys and apes. Note the upward shift in position of the transverse process in ape lumbar vertebrae relative to that of monkeys.

shape and orientation as well, with perpendicularly oriented spinous processes.

Oddities

Lorises have broad ribs in comparison to those of other primates. Lorises are also known for their dorsoventral foramina at the base of the transverse processes along their thoracic vertebrae. Lorisids have short lumbar regions, especially the large lorises *Nycticebus* and *Perodicticus*, as well as short spinous processes. Lorisids require lumbar stability to counter the bending stresses that occur during their

anti-pronograde and flexible back postures. Lorisid movements have often been described as serpentine. Tarsiers have highly bent ribs and a more barrel-shaped thorax compared to those of other pronograde primates.

THE PELVIS AND PRIMATE REPRODUCTION

The primate pelvis, or hipbone, is a locomotor structure, a reproductive region, and an intestine container. It is a rigid structure made up of several bony elements and it attaches to the vertebral column on the right and left sides of the body (see fig. 7.2). A fused innominate bone is actually composed of three bones: the ilium, the ischium, and the pubis. The sacrum, holding the vertebral column above, attaches to the right and left innominate, respectively, while the pubic symphysis is bound together by fibrocartilage. Pelvic shape changes across primates will not be addressed here but is discussed in chapter 9. This section focuses on the reproductive implications for the pelvic region relative to the vertebral column.

In terms of reproduction, pubic bones are longer and thinner across primates, but no real functional distinctions have been ascribed to these morphological lengths. For most primates, the birth canal is quite wide relative to the size of the neonate head, even given the relatively large brain sizes of primates. In contrast, neonate head size is equal to or exceeds the width of the birth canal opening in *Saimiri*, *Cebus*, macaques, and humans. This reduction in birthing space causes difficulties in delivery, and in humans a series of baby rotations need to occur before crowning.

The vertebral column of primates houses and protects many of the internal organs of the body—from the heart and lungs within the thorax to the reproductive structures within the pelvis. When primates become pregnant and embryos begin to develop, it is the placenta that allows for the exchange of nutrients and excretory materials for early growth and development. Primate placentas come in two varieties to protect relatively long-term babies. Strepsirhine primates possess an epitheliochorial placenta (fig. 7.14) in which the capillaries of the mother's tissue are separated from the placenta and embryo by an epithelial cell lining within the uterus. This cellular lining prevents close contact with the mother's tissues and blood system, thereby providing a slower transfer of nutrients and exchange of excretion materials. After birth, however, the epitheliochorial placenta separates from the uterine wall without any maternal tissue and little bleeding occurs. In contrast, haplorhine primates possess a haemochorial placenta with capillaries that burrow into the uterine wall making direct contact (i.e., placental discs) with maternal tissues and maternal blood flow (fig. 7.14). The haemochorial placenta has a greater transfer of nutrients but tears away at birth with extensive blood and tissue loss from the mother. The haemochorial placenta, however, offers a functional advantage for the bigger-brained haplorhine fetus in terms of efficient nutrient access during early growth and development.

Primates can also be distinguished by two types of uterine shapes. Strepsirhine primates and tarsiers have a bicornuate uterus, while anthropoids have a unicor-

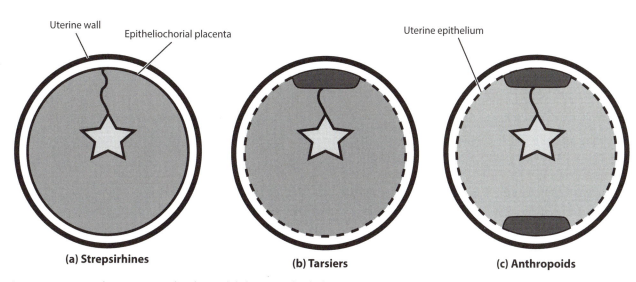

Figure 7.14 Primate placentas. Note the placental disks among haplorhine primates.

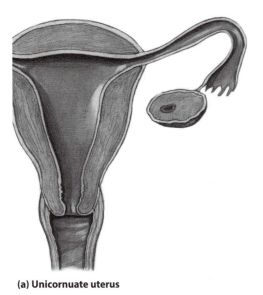

(a) Unicornuate uterus

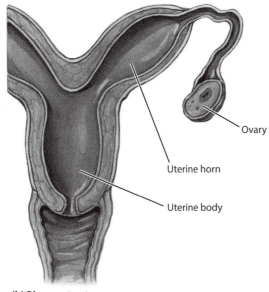

Ovary

Uterine horn

Uterine body

(b) Bicornuate uterus

Figure 7.15 Two forms of the primate uterus.

nuate uterus. A bicornuate uterus has a small uterine body with larger uterine horns (fig. 7.15), whereas an unicornuate uterus has a large uterine body and virtually no uterine horns. Tarsiers are intermediate in uterine structure, having a large body and large uterine horns (fig. 7.15). The advantage of these two uterine types is tied to the number of neonates that species commonly give birth to. Strepsirhine species that commonly birth twins utilize the bicornuate uterus with the two horned chambers, while haplorhines normally birth one large-brained infant

that grows within the large body of the unicornuate uterus. Of course, there are interesting exceptions like the callitrichine monkeys that commonly birth twins while possessing an unicornuate uterus.

Two other soft tissues in primates are involved in lactation and reproduction: mammary glands and penises. Mammary glands, nipples or teats, are distributed in different locations along the body of primates. In dwarf lemurs, lorises, galagos, and tarsiers, one or two pairs are located on the abdomen. In aye-ayes they are inguinal. Among the New World monkeys, nipples are located near the arm pits, while catarrhines and lemurs have pectoral nipples. With the exception of humans, none are relatively enlarged. Suckling is feeding on demand by primate infants. Primate mothers and infants remain in close contact for most of early infant development. Only a few primates (e.g., galagos, tarsiers, and cheirogaleids) park their infants to go off and feed. Milk composition is quite watery across primates and low in fat and protein content.

In terms of penis structure or testicle size, most primates have a short and simple penile structure with a baculum. Baculums are small penis bones that are elongated in lorises, galagos, dwarf lemurs, and the aye-aye, primates that have dispersed mating systems and multiple partners. Similarly, penile spine development is best developed among these same dispersed mating prosimians. Baculums are reduced in great apes and absent in humans, tarsiers, and several atelines. Human males display an elongated penis relative to that of the other great apes.

Testicle size is reflected in sperm volume. Male primates that participate in dispersed mating systems or ones that have promiscuous mating patterns have increased testicle size and sperm volume to compete against other males. For example, male chimpanzees are known to possess enlarged testicles, and many strepsirhines undergo seasonal changes in testicle size.

Oddities

Perhaps the oddest morphological features concerning reproduction and lactation are the excessive sexual swellings of female chimpanzees and bonobos (see chapter 3) or the enlarged breasts of female humans relative to those of the great apes.

Selected References

Anderson, J.E. 1983. Grant's Atlas of Anatomy. 8th edition. Williams and Wilkins, Baltimore.

Ankel-Simons, F. 2000. Primate Anatomy—An Introduction. 2nd edition. Academic Press, New York.

Ankel-Simons, F., and C. Simons. 2003. The axial skeleton of primates: how does genus *Tarsius* fit?; pp. 121–144 *in* P.C. Wright, E.L. Simons, and S. Gursky (eds.), Tarsiers—Past, Present and Future. Rutgers University Press, New Brunswick, N.J.

Basmajian, J.V. Primary Anatomy. 7th edition. Williams and Wilkins, Baltimore.

Cartmill, M., and K. Milton. 1977. The lorisiform wrist and the evolution of "brachiating" adaptations in the Hominoidea. American Journal of Physical Anthropology 47:249–272.

Dixson, A.F. 1998. Primate Sexuality: Comparative Studies of the Prosimians, Monkeys, Apes, and Human Beings. Oxford University Press, Oxford, UK.

Erikson, G.E. 1963. Brachiation in New World monkeys and in anthropoid apes. Symposium of the Zoological Society of London 10:135–164.

Fleagle, J.G., and F. Anapol. 1992 The indriid ischium and the hominid hip. Journal of Human Evolution 22:285–305.

Fleagle, J.G., and E.L. Simons. 1979. Anatomy of the bony pelvis in parapithecid primates. Folia Primatologica 31:176–186.

Gebo, D.L., M. Dagosto, K.C. Beard, and X. Ni. 2008. New primate hind limb elements from the Middle Eocene of China. Journal of Human Evolution 55:999–1014.

Hartman, C.G., and W.L. Straus (eds.). 1971. The Anatomy of the Rhesus Monkey. Hafner Publishing Company, New York.

Hildebrand, M. 1959. Motions of the running cheetah and horse. Journal of Mammalogy 40:481–495.

Hurov, J. 1987. Terrestrial locomotion and back anatomy in vervets (*Cercopithecus aethiops*) and patas monkeys (*Erythrocebus patas*). American Journal of Primatology 13:297–311.

Jenkins, F.A. 1970. Anatomy and function of expanded ribs in certain edentates and primates. Journal of Mammalogy 51:288–301.

Johnson, S.E., and L.J. Shapiro. 1998. Positional behavior and vertebral morphology in atelines and cebines. American Journal of Physical Anthropology 105:333–354.

Kapandji, I.A. 1974. The Physiology of Joints, Vol. 3, The Trunk and Vertebral Columns. Churchill Livingstone, London.

Keith, A. 1923. Man's posture: its evolution and disorders. British Medical Journal 1:451–454, 545–548, 587–590, 624–626, 669–672.

Le Gros Clark, W.E. 1959. The Antecedents of Man: An Introduction to the Evolution of Primates. University of Edinburgh Press, Edinburgh.

Luckett, W.P. 1993. Developmental evidence from the fetal membranes for assessing Archontan relationships; pp. 149–186 *in* R.D.E. MacPhee (ed.), Primates and their Relatives in Phylogenetic Perspective. Plenum Press, New York.

Martin, R.D. 1990. Primate Origins and Evolution—A Phylogenetic Reconstruction. Princeton University Press, Princeton, N.J.

Niemitz, C. 1984. Locomotion and posture of *Tarsius bancanus*; pp. 191–225 in C. Niemitz (ed.), Biology of Tarsiers. Gustav-Fischer, Stuttgart, Germany.

Rose, M.D. 1975. Functional proportions of primate lumbar vertebral bodies. Journal of Human Evolution 4:21–38.

Schultz, A.H. 1969. The Life of Primates. Weidenfeld and Nicolson, London.

Shapiro, L. 1993. Functional Morphology of the Vertebral Column in Primates; pp. 121–149 *in* D.L. Gebo (ed.), Postcranial Adaptation in Nonhuman Primates. Northern Illinois University Press, DeKalb.

Shapiro, L., and C.V.M. Simons. 2002. Functional aspects of strepsirhine lumbar vertebral bodies and spinous processes. Journal of Human Evolution 42:753–783.

Tattersall, I. 1982. The Primates of Madagascar. Columbia University Press, New York.

Walker, A.C. 1970. Nuchal adaptations in *Perodicticus potto*. Primates 11:135–144.

Ward, C.V. 1993. Torso morphology and locomotion in *Proconsul nyanzae*. American Journal of Physical Anthropology 92:291–328.

Washburn, S.L. 1968. The Study of Human Evolution (Congdon Lectures). University of Oregon Books, Eugene.

Primate Locomotion and the Forelimb

PRIMATE LOCOMOTION

Primates generally move through trees given their many arboreal adaptations. They commonly prefer above-branch quadrupedalism, but they climb up and down vertical supports, and leap or swing across spatial gaps (fig. 8.1). Compared to other mammalian Orders, primates display a high variety of movement patterns. Some, like the mouse lemurs and lorises, utilize below- or side-branch arboreal quadrupedalism (suspensory quadrupedalism), while the apes and spider monkeys use forelimb suspension and brachiation. Other primates have come to the ground as terrestrial quadrupeds but are still capable of climbing and moving within trees. Human bipedalism is undoubtedly the oddest movement pattern used to traverse a terrestrial environment.

Grasping hands and feet are the hallmark adaptation for the entire Order, and all living primates but humans possess a grasping big toe (the hallux). Primate locomotion is normally divided into four broad locomotor categories: quadrupedalism, leaping, climbing, and suspensory behaviors, usually with the hindlimbs but including forelimb suspension and brachiation in the apes and spider monkeys.

Quadrupedalism, either arboreal or terrestrial, fast (running) or slow (walking), shows primates to be hindlimb driven with greater propulsive force being applied by the hindlimbs rather than being evenly distributed across each of the four limbs. It appears that primates actively engage hindlimb muscles, hip

extensors, to shift their body weight back onto their hindlimbs. Other quadrupedal mammals, like horses or dogs, are forelimb dominated, the opposite of primates. The initial ancestral use of the forelimb and hand as a strictly weight-bearing structure changed within the evolutionary sequence of primates. Primates use their forelimbs and hands for reaching, foraging, feeding, and, of course, grasping. In fact, primate ancestry has often played a crucial role in the evolution and subsequent adaptive locomotor modes of primates.

Arboreal quadrupedalism in primates differs from that of other mammalian quadrupeds in that primates use a diagonal couplet gait sequence. This allows one limb to move forward while the other three grasp a support, a cautious and safe strategy for animals high up in the canopy. The sequence goes like this: left hand, right leg, right hand, and left leg (fig. 8.2). The unique nature of primate quadrupedalism is likely due to their grasping appendages. Primates move with a protracted humeral position, allowing a greater forelimb stride relative to that of other quadrupedal mammals. Primates also bend their elbows and knees to move their torsos closer to a support when grasping and moving along a branch. We call this gait compliance. Gait compliance helps keep a primate's body mass balanced over an unstable arboreal support. Life in the canopy requires a constant awareness to keep from tipping over the edge of a support and primates are designed to do just that.

Quadrupedal primates tend to have more equal

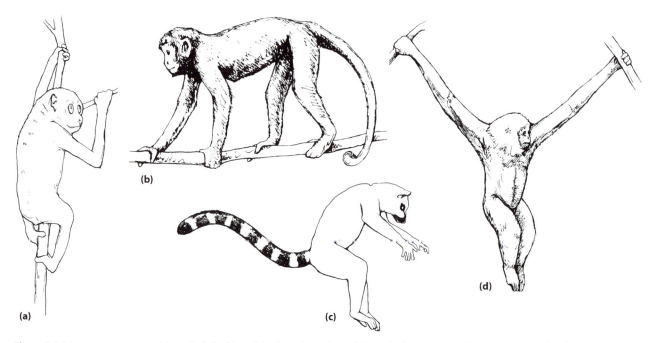

Figure 8.1 Primate movements: (a) vertical climbing; (b) arboreal quadrupedalism; (c) leaping; and (d) arm swinging (brachiation).

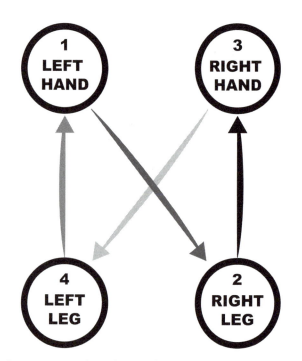

Figure 8.2 Diagonal couplet gait of primates.

limb lengths for gait efficiency. Consider the problem we humans have when we attempt to revert back to quadrupedalism. Our legs are far too long, relative to our arms, for us to be efficient at moving quadrupedally—not to mention keeping our heads looking forward. Normally, we cheat and move along the ground quadrupedally on our knees with our entire lower limbs touching the ground. We can move this way, but we are not very good at it (e.g., there is no way we can outrun predators).

Longer limbs with good joint mobility, a characteristic of primates, provide longer strides and a more efficient movement pattern. Consider the distance a horse's limb moves per stride relative to that of a pig's. The longer the distance or race, the less muscle contractions will be needed to move a longer limb across the same distance. Short limbs need more strides and require more muscle contractions to cover the same distance, allowing muscle fatigue to build up over the same time interval. In short, long-limbed animals outrun short-limbed ones, especially as the distance to traverse increases.

Quadrupedal primates pronate their hands, meaning that they turn them inward to grab hold of an arboreal support with the palm's surface touching the curved surface of a branch (fig. 8.3). We call this a palmigrade hand position. Whole hand contact with the ground is also a palmigrade hand posture. By extending the digits, digitigrade hand postures can be utilized by several terrestrial primates, thus elongating the forelimb as well (fig. 8.3). Orangutans tend to fold their fingers and use a fist-walking technique when moving quadrupedally along the ground. In

condyle, similar fore and hindlimb limb lengths, and short, broad carpal bones. They sacrifice mobile joints for joint stability.

Leaping is one of the extraordinary movements of primates. Leaping allows primates to cross spatial gaps quickly in the canopy. It is efficient in the sense that primates do not need to double-back to re-cross a distance already traveled to move ahead. Leaping allows tree-to-tree crossing and it is a fast movement that puts a considerable distance between a primate and a predator. On the other hand, falling is dangerous, particularly as primates increase in size. As you might remember from chapter 4, bone strength is measured by its cross-sectional area (length squared) relative to body mass, which is a cubed variable (length to the third power). As a result, size increases more steeply relative to bone strength, leaving larger animals at greater risk of falling.

Primates leap in two ways: forceful upward parabolic leaps and horizontal downward leaps (fig. 8.4). Strepsirhine primates like galagos, lemurs, *Lepilemur*, and indriids are especially known for their forceful and propulsive upward leaping styles, as are the haplorhine tarsiers. Horizontal downward leaps characterize anthropoid primates, although the callitrichids seem to be able to use both types. Forceful leaping primates have long legs and they propel themselves upward over long distances with a parabolic trail of movement. They tend to tuck their legs and arms at mid-leap. These leapers are notorious for the length of their leaps, their precision, and their accuracy upon landing. Many of these leapers are described as vertical clingers and leapers (i.e., indriids, *Lepilemur*, many galagos, and tarsiers), being able to vertically cling to a vertical branch and leap across to another vertical support. Vertical clingers and leapers propel themselves backward, or sideways, and away from their initial substrate, twisting in mid-air, and are still able to face the oncoming landing support, usually contacting it feet first. They are capable of propelling themselves across distances many times their small body lengths.

Takeoff speed is one critical variable relative to the distance covered. The distance of a leap depends on takeoff velocity and the angle of takeoff (the best angle equals 45 degrees). Forceful hindlimb propulsion allows a distance to be crossed quickly via leaping. Long legs produce leaps with less relative muscle force

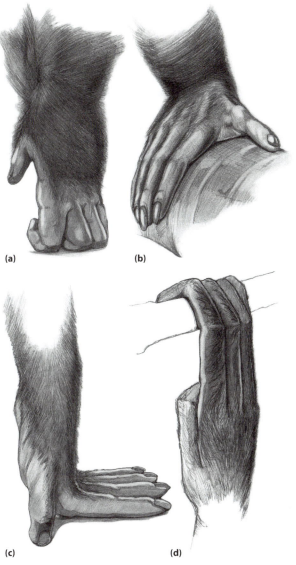

Figure 8.3 Hand positions: (a) a knuckle-walking hand position; (b) a palmigrade hand position; (c) a digitigrade hand position; and (d) a suspensory grasp.

contrast, African apes bend their fingers and walk on their knuckles (i.e., knuckle walking; fig. 8.3). The great apes clearly have fingers that are too long for an efficient quadrupedal gait and all are attempting to compensate for this mechanical deficiency with these odd finger-folded hand positions. The long arms of apes, relative to their legs, with shoulders elevated well above their hips, hinder their quadrupedal efficiency as well. With perhaps the exception of patas monkeys, primates are not particularly good cursors akin to horses or dogs. Terrestrial monkeys tend to have relatively shorter digits, restricted shoulder mobility, narrow elbows with a posteriorly oriented medial epi-

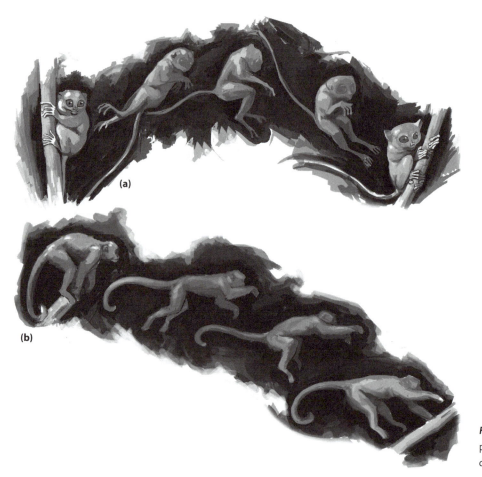

Figure 8.4 Leaping styles: (a) a parabolic leap and (b) a horizontal downward leap.

in contrast to short-legged primate leaps, more often performed by anthropoid primates.

Anthropoid leapers can be quite frequent but they tend to leap outward horizontally, thereby falling downward. Anthropoid leapers rarely land higher or at the same height as their initial takeoff position. They lack the relatively large thigh muscles and longer leg mechanics of more forceful leaping primates. Anthropoids are also a contrast in precision. They are often observed crashing through terminal branches after long downward leaps before catching hold of branches and eventually slowing their descent and coming to a final stop. Anthropoid leaps are more efficient when crossing from one horizontal branch to another. Regardless of the lack of beauty, many anthropoids leap frequently and are quite adept at this movement.

In terms of leaping and landing, primates can land feet first with their hindlimbs as the initial shock absorbers, use their hands or forelimbs first (e.g., *Euoticus* or *Galago alleni*), or use all four limbs to initially con-

tact a support. Landing forces may exceed 10 times an individual leaper's body weight.

Grasping and climbing go hand in hand in primate evolution. Grasping and climbing are the initial movement patterns for all primates. Grasping is the key adaptation that allows arboreal primates access to the small-branch microhabitat. Vertical climbing allows primates to ascend vertical supports to position themselves at different canopy heights, an obvious advantage while foraging or escaping danger. Climbing is often divided into vertical climbing and more oblique climbing, often called clambering. This division reflects the physical difference in ascending a vertical tree trunk or a liana versus a more quadrumanual progression through tangles and oblique branch angles. Climbing requires good joint mobility and flexible limbs, besides grasping hands and feet. Grasping is an advantage when moving along small supports, like terminal branches. In contrast, claws and claw climbing are better suited for large-diameter supports like tree trunks. Primates that often use large-diameter vertical supports (e.g., callitrichids, *Phaner*, or *Euoticus*) have re-evolved claw-like nails to offset their grasping mechanics, which is inefficient on relatively large-diameter surfaces. The effort to resist the pull of gravity relative to body mass is more difficult as a primate climbs upward via grasping on an ever increasing support diameter. As tree girth increases, these surfaces ultimately become a relatively flat wall to primates and a climbing primate can no longer apply enough force to maintain a hold on these large-diameter supports. Falling ensues. To avoid this dangerous situation, primates often climb lianas or vines, small-diameter supports, along the sides of large tree trunks. Even terrestrially adapted primates (e.g., baboons) retain the ability to climb trees. Only humans (with perhaps the exception of children) have given up on their tree-climbing past.

Some primatologists like to view bridging as a climbing-related movement pattern. In bridging primates extend their bodies horizontally across short gaps and bridge across to the other side (fig. 8.5).

Bridging is a different type of movement pattern that simply helps primates cross horizontal gaps in the canopy. It rarely is involved in upward progression as in climbing and is usually utilized by primates that do not leap.

Suspensory locomotion can entail inverted quadrupedal walking where the limbs are in tension or arm hanging and brachiation. Small primates like cheirogaleids or lorises often use inverted quadrupedal movements underneath or along the sides of branches. They can be quick, slow, or stealthy when moving in this manner.

Brachiation is a common movement pattern for gibbons and for the spider and woolly spider monkeys (fig. 8.6). All of the living apes, including humans, can perform this arm-rotational behavior, as can some colobine monkeys and *Lagothrix*. Overall, frequent brachiation is a fairly rare movement pattern, being confined to the clades of living apes and the spider monkeys. Frequent brachiators, like the living apes, have also fundamentally changed the anatomy of their forelimb and thorax to accomplish this motion. The

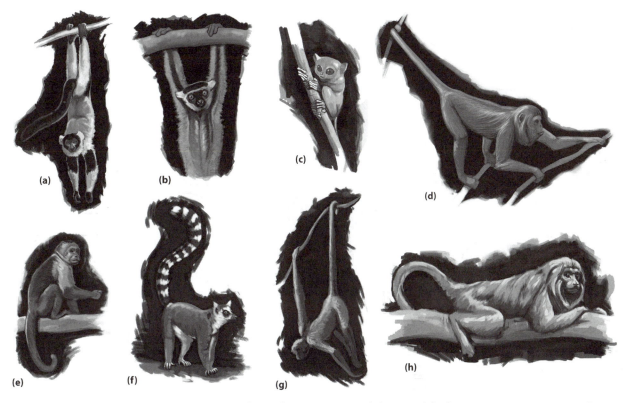

Figure 8.5 Primate postures: (a) hindfoot suspension; (b) arm hanging; (c) vertical clinging; (d) bridging (a gap-crossing maneuver); (e) sitting; (f) standing; (g) tail and hand suspension; and (h) lying down.

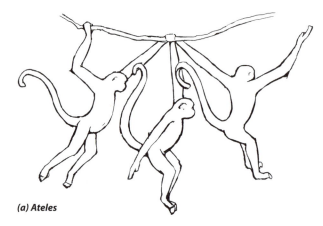

(a) Ateles

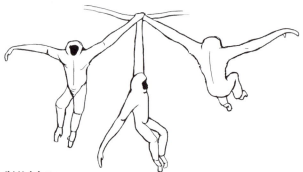

(b) Hylobates

Figure 8.6 Brachiation: (a) a spider monkey with an oblique trunk orientation and tail-assisted grasping arm swing relative to a gibbon (b), with its elongated arms and a vertical trunk orientation.

large New World monkeys, the atelines, have added a fifth limb, a prehensile tail that aids in grasping branches while brachiating, in contrast to the tail-less apes, and atelines do not extend nor bend their knees while brachiating as do the apes.

POSTURES

Besides movements, primates also utilize postures. Primates stand, sit, or lie down often (fig. 8.5). Hindlimb suspension is also a common suspensory posture across a variety of primates, often used during feeding. The atelines often use this posture in conjunction with their prehensile tail to obtain food items. Arm hanging is a much rarer suspensory posture across primates, confined generally to spider monkeys and the Asian apes.

One of the most interesting primate postures is cantilevering (fig. 8.7). Here primates like galagos, lo-

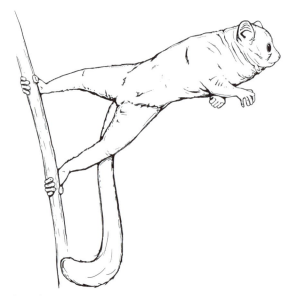

Microcebus

Figure 8.7 Cantilevering by *Microcebus*.

rises, tarsiers, and cheirogaleids grasp vertical supports with their feet and extend their bodies horizontally outward. This enables primates to use their hands to grasp insects while only holding on to a support with their hindlimbs. This is a truly impressive maneuver considering the strength and precision required to extend a body horizontally away from a vertical support.

LIMBS

Primate limbs are generally quite long relative to those of other mammals of equal size. This is especially true for the long-legged leaping primates or the long-armed brachiating primates. Three limb indexes are helpful in examining primate limb lengths: the intermembral index, the brachial index, and the crural index. The intermembral index examines the lengths of the four major long bones: the humerus and the radius relative to the femur and the tibia. A high intermembral index (> 100) implies a long forelimb and is associated with brachiating primates (e.g., gibbons). A low intermembral index (< 70) implies a long leg and is associated with leaping primates (e.g., tarsiers) or human bipedalism. Primates with intermembral indexes closer to 100 have forelimbs and hindlimbs of about equal length and these primates tend to be quadrupedally oriented (e.g., baboons). The intermembral index also increases as body size increases.

The brachial index examines radial length relative to humeral length or forearm to upper arm proportions. Apes, especially gibbons, have a long humerus and even longer forelimb bones (radius and ulna) with a brachial index above 110. *Loris* has a long forearm at 115 for this index as well. Many primates have an index above 100, however (e.g., *Ateles, Avahi, Erythrocebus, Galago, Hapalemur, Indri, Eulemur, Lepilemur, Mandrillus, Papio, Perodicticus, Presbytis, Pygathrix, Tarsius,* and *Theropithecus*). This list combines brachiators, cautious-moving lorises, terrestrial baboons, and leaping tarsiers together with no assortment by locomotor mode.

The crural index examines the same pattern of limb lengths as the brachial index but with the lower limb (tibia length relative to femur length × 100). Many different locomotor modes can have a similar crural index. For example, *Nycticebus coucang*, a non-leaping loris, has a value of 101 while *Tarsius bancanus*, a frequent leaping primate, has a value of 100. Many indriids and lemurs have values below 100, indicating a longer femur than tibia. In fact, most primates are at or more often below 100 in this index. Often smaller taxa, like *Microcebus*, possess longer tibiae relative to their femora.

Comparing forelimb and hindlimb relative length proportions across primates shows taxa like the slender lorises (*Loris* and *Arctocebus*) to have exceptionally long forelimbs and hindlimbs. Cheirogaleids are known for the relatively shortest forelimbs and hindlimbs relative to other primates. This is surprising given the small body sizes of this family and the fact that many are good leapers. Gorillas also have a very short hindlimb, while lemurs have longer hindlimbs than forelimbs, a characteristic of most leaping primates. Tarsiers have the relatively longest hindlimbs with a pattern similar to that observed within the galagos and to a lesser extent with indriids. Indriids have long femora relative to their tibias. Brachiators like gibbons, orangutans, and spider monkeys are known for their elongated forearms. *Daubentonia* and *Tarsius* are known for their long hands and long fingers, while patas monkeys, gelada baboons, and savanna baboons are known for their relatively short digits.

Humeral shafts are generally straight as in the apes and forelimb suspensory primates or retroflexed (bent anteroposteriorly; fig. 8.8) as in quadrupedal primates. Humeral shafts are longer in some primates and this

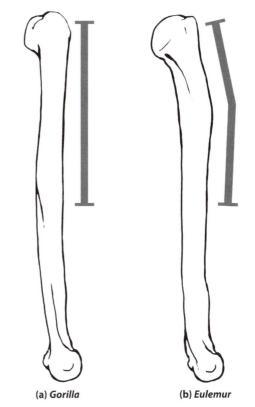

(a) *Gorilla*   **(b) *Eulemur***

Figure 8.8 Humeral shaft orientation: (a) straight (in apes, *Gorilla*) and (b) retroflexed (in most primates, *Eulemur*).

can be examined with the brachial index or relative segmental lengths of the forelimb. Humeral cortical thickness increases for the slow-climbing primates, like lorises, indicating an adaptation to higher axial loads on the humerus.

Femoral shafts are generally round but can be oval-shaped in the anteroposteriorly (a-p) dimension or flattened and widened in the mediolateral (m-l) dimension. Lorises, for example, show high m-l/a-p ratios, while the leaping indriids show the opposite pattern. Cortical bone thickness increases in the slow-climbing and powerful gripping taxa. Rapid locomoting primates are dominated by a-p bending and this is reflected in higher a-p/m-l ratios.

FORELIMB
Shoulders

The shoulder joint falls into two functional varieties among primates: type 1—parasagittal limb movements (generally fore and aft arm movements); and type 2—circumduction (an around-the-world arm

Figure 8.9 Arm motions: (a) a parasagittal shoulder movement (largely a fore and aft movement of the arm relative to circumduction); (b) an around-the-world shoulder movement.

movement) (fig. 8.9). The parasagittal shoulder joint is common among primates, while circumduction is found only in the living apes (including humans) and spider monkeys (*Ateles* and *Brachyteles*). In terms of anatomical structure, the parasagittal shoulder has a posteriorly facing humeral head with good-sized tubercles, the greater and lesser tubercles, along the sides of the humeral head. In some taxa, the humeral head extends above the tubercles (e.g., lorises, galagids, indriids, *Lepilemur*, tarsiids, atelines, some colobines, hylobatids, and pongids), is flattened dorsally (terrestrial quadrupeds like Old World monkeys), is equal in height with the greater tubercle (e.g., lemurs, most platyrrhines, and some colobines), or is positioned below the tubercles (terrestrial cercopithecids).

In parasagittal shoulder joints the glenoid joint facet of the scapula is along the side of the narrow thorax and thus the shoulder is close to the midline of the body (see fig. 7.1). Clavicles are short in these primates and the scapula is long from the glenoid facet to the vertebral border (fig. 8.10). The parasagittal shoulder joint allows for fore and aft movement with limited mediolateral mobility, although mediolateral shoulder mobility in primates is greater than that of other mammals.

When the humerus is flexed or retracted, the bottom half of the humeral head articulates with the glenoid facet of the scapula. When the humerus moves forward into extension, the top half articulates with the glenoid facet. The humeral head stops its movement at the extremes of the fore and aft humeral positions; many refer to this joint situation as close packed (fig. 8.11). Taxa with larger and taller greater tubercles (i.e., Old World monkeys) have less shoulder mobility than those with smaller tubercles below the humeral head since a taller greater tubercle contacts the glenoid facet sooner, preventing any further joint motion. Terrestrial quadrupeds, like Old World monkeys, have

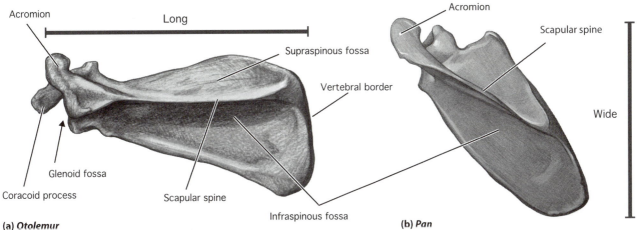

(a) Otolemur

(b) Pan

Figure 8.10 Scapulae: (a) the long scapula common among most primates; and (b) the wide scapula of living apes.

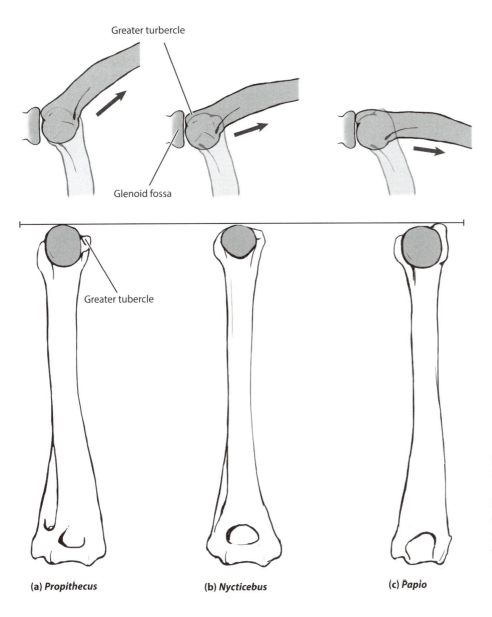

(a) Propithecus

(b) Nycticebus

(c) Papio

Figure 8.11 Humeral head height relative to the greater tubercle: (a) an indriid with a tall humeral head relative to the greater tubercle; (b) a loris with a humeral head about equal to the greater tubercle; and (c) an Old World monkey with a tall greater tubercle relative to the humeral head.

flattened the top of the humeral head, and this shape change adds stability during terrestrial quadrupedalism. The humeral shaft is quite retroflexed in these taxa as well.

For the living apes, the shoulder is highly mobile and capable of making a full circle above the head (circumduction; see fig. 8.9). This movement pattern necessitates a host of morphological changes from that of a monkey, a tarsier, or any strepsirhine primate. First, the humeral head is a large, ball-like structure that provides considerable joint mobility (fig. 8.12). A large, round joint surface is coupled with smaller greater and lesser tubercles that are located close together. More importantly, the humeral head of apes faces medially toward the body rather than posteriorly as in quadrupedal monkeys and other primates. This twisting of the humeral head relative to that of the humeral shaft is called medial torsion (fig. 8.12). Medial torsion is higher among the great apes, especially African apes and humans, relative to gibbons. Medial torsion allows the humeral head to contact the round and oval glenoid joint surface of the scapula, which

faces laterally now as it lies along the back of the dorsally flattened thorax of apes, as well as positioning the distal humerus or elbow forward. Most primates possess a more elliptically shaped glenoid fossa in contrast to the more expansive joint for ape shoulders. Apes also need long clavicles since the shoulder of apes is pushed away from the midline, located on the sides of the body, while still contacting the reoriented scapula lying on the back of the thorax (see fig. 7.10). A large acromial process on the scapula helps enclose this shoulder joint region and provides a critical anchor for the large shoulder muscles of apes, especially the deltoid muscle. Likewise, the scapula of apes is long craniocaudally (fig. 8.10), versus being long from the glenoid facet to the vertebral border, providing the attachment surface for many shoulder muscles related to arm positions over the head (i.e., the rotator cuff: the supraspinatus, the infraspinatus, the teres minor, and the subscapularis). The deltoid muscle helps elevate and abduct the humerus with the supraspinatus, while the lower rotator cuff muscles abduct and laterally rotate the humerus (the infraspinatus), abduct

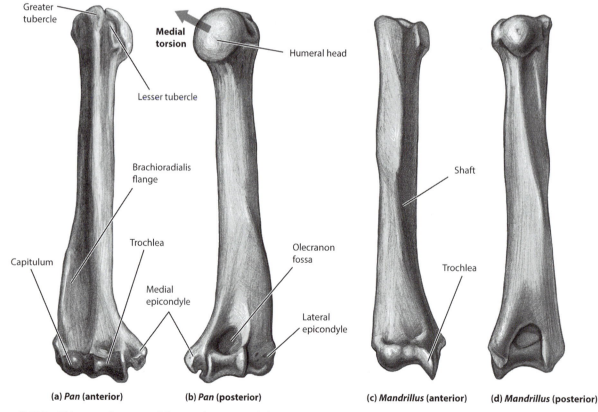

(a) Pan (anterior)   **(b) Pan (posterior)**   **(c) Mandrillus (anterior)**   **(d) Mandrillus (posterior)**

Figure 8.12 An African ape humerus (*left, Pan*) showing medial torsion of the humeral head; and (*right*) the humerus of a mandrill (*Mandrilllus*).

and medially rotate (the subscapularis, top and middle portions), adduct and medially rotate (lower part of the subscapularis), and adduct and laterally rotate (the teres minor) the humerus (see chapter 4).

ODDITIES

Lorises have expanded their humeral heads into a large, ball-like joint structure similar to that of apes and have reduced their greater and lesser tubercles, although their humeral heads still face posteriorly like other parasagittally oriented primate shoulders (see fig. 8.11). Although lorises are not brachiators, these structural modifications do allow them to increase shoulder mobility; in fact, they are known for their highly flexible body joints and for their serpentine movement patterns.

Indriids have tall and narrower humeral heads that extend above the greater tubercle (see fig. 8.11). This humeral head shape is unusual among other living primates.

Lagothrix is an interesting New World monkey since it possesses a humeral head similar to the parasagittal forms; yet, it is capable of performing brachiating movements. *Lagothrix* does possess slightly greater medial torsion, about 10 degrees greater, than most other primates as well as a few structural modifications within its thorax. Brachiation with a parasagittal shoulder morphology is interesting since *Lagothrix* may bridge an intermediate evolutionary position

prior to frequent brachiation in spider monkeys (*Ateles* and *Brachyteles*). Other large colobines (e.g., *Pygathrix*) also seem to show a similar morphology intermediate pattern for their shoulder anatomy.

Elbows

The distal humerus is another area in the primate body where joints and muscle attachment surfaces are in close contact to each other. The distal humerus holds a key joint complex across primates in that primate forearm and hand mobility is often uniquely tied to their ancestral arboreal and feeding adaptations. At the distal humerus, the capitulum and trochlea (fig. 8.13), the head of the radius, a circular joint surface with a central depression, and the ulnar trochlear notch, as well as the ulnar olecranon process are all key bony structures that make up the primate elbow joint complex. In non-anthropoid primates, the capitulum, the joint surface for the radial head, is round and ball-like, and this structure is separated from the trochlea, the joint surface for the ulna, by a wide groove or gutter called the zona conoidea (fig. 8.13). In strepsirhines and tarsiers, the capitulum joint surface often has a capitular tail, a small joint surface that extends away from the capitulum and toward the lateral epicondyle.

All of the hand rotations that primates, and humans, are capable of occur at the elbow joint com-

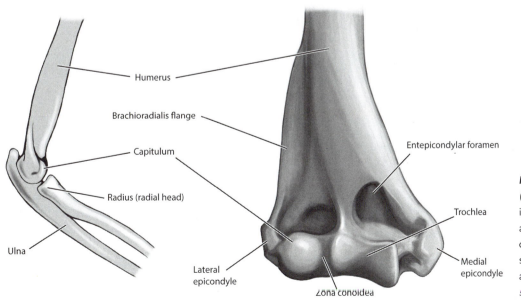

Figure 8.13 Primate (*Eulemur*) distal humerus illustrating elbow joint anatomy (i.e., the capitulum, the joint surface for the radius, and the trochlea, the joint surface for the ulna).

Humerus

Brachioradialis flange

Capitulum

Radius (radial head)

Ulna

Lateral epicondyle

Zona conoidea

Entepicondylar foramen

Trochlea

Medial epicondyle

We often describe fossil primates as being similar to living primates, but these statements and comparisons are only true to a point. Our generalizations offer a helpful frame of reference and imply an adaptive or ecological model to envelop an ancient primate. Consider as an example the fossil primates from the Eocene called adapiforms, often described as lemur-like primates. The body size, diet, and movement patterns of adapiforms were similar to those of lemurs. The box on page 86 showed how the tooth formula of *Notharctus tenebrosus* was not lemur-like in terms of premolar numbers and that *Notharctus* did not possess a toothcomb. Illustrated below is another example of anatomical differences between the Eocene fossil *Notharctus* and living lemurs. On the left is an anterior view of the humerus of *Notharctus* and on the right the humerus of *Eulemur fulvus*. Note the extensive development of the brachioradialis flange and the deltopectoral ridge in *Notharctus* relative to the humerus of *Eulemur*. Although both possess similarly working shoulder and elbow joints, *Notharctus* had more prominent muscle attachment sites, implying better-developed arm muscles relative to those of living lemurs.

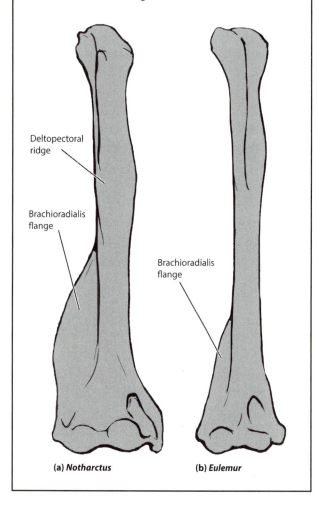

Deltopectoral ridge

Brachioradialis flange

Brachioradialis flange

(a) *Notharctus*   **(b) *Eulemur***

plex, not at the wrist joint. These considerable and impressive rotational movements, hand supination and pronation, occur between the radius, actually the radial head, and the capitulum of the humerus in primates. This joint is responsible for primate forelimb rotations. The elbow joint and subsequent forelimb movements are often associated with frequent climbing and grasping activities. Palmigrade grasping hands are in a supinated forelimb position as primates move through the canopy.

The trochlea of the humerus functions with the ulnar notch in flexing and extending the forearm. The trochlear surface can be tall and narrow, as in vertical clinging and leaping primates (fig. 8.14), or short and wide. In all primates but the living apes, the ulna possesses a moderate to long ulnar process, the olecranon process, and the attachment site of the triceps muscle (fig. 8.15). This bony process prevents full extension in the forearm of most primates so that most primate elbows are bent (flexed) since they cannot be fully extended; this is not the case, however, in apes and humans.

The distal humerus in non-anthropoid primates tends to be wide from the medial to lateral epicondyles with prominent epicondylar surfaces and a large bony extension proximally along the lateral edge, the brachioradialis flange, the attachment site for the brachialis and brachioradialis muscles (see fig. 8.13). The brachialis muscle is important in the rotational movements of the elbow and hand. Non-anthropoid elbow regions tend to be muscular and most possess an entepicondylar foramen on the medial side as well. The brachioradialis flange is prominent in *Daubentonia* and large in its expansion in cheirogaleids, galagos, tarsiers, *Lepilemur*, and lemurs. Indriids and lorises have reduced this structure, as have anthropoid primates.

Anthropoid elbow morphology tends to have a less rounded or more cylindrical-shaped capitulae (fig. 8.14), and anthropoids have lost or greatly reduced the space between the capitulum and the trochlea (i.e., the zona conoidea). In anthropoids, the trochlea is often like a slanting cylinder. The trochlea is short in height and wide in mediolateral length. The distal humerus is narrower in anthropoids with little or no brachial flanges, limiting muscle attachment areas. These elbow features provide stability during flexion and extension with a fully pronated forearm during

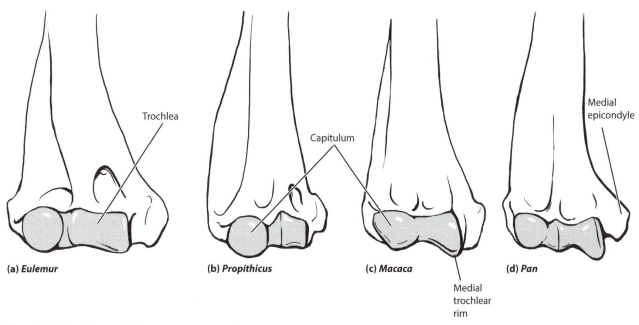

Figure 8.14 Distal humeral joint anatomy across primates.

quadrupedalism. In Old World monkeys the medial rim of the trochlea is greatly pronounced (fig. 8.14), a feature associated with elbow support for terrestrial movements.

In the living apes, the capitulum is ball-like and the trochlea is beveled (see fig. 8.14). This beveling, a trochlear notch, allows the ulna to be held firmly on the trochlea of the humerus during full extension of the forearm by the living apes. The olecranon process is virtually eliminated among living apes (fig. 8.15); thus it does not hinder full extension of the forearm as it does among other primates. Overall, living apes have mobile elbows as well as shoulders to perform brachiation, arm suspension, and climbing movements; this elbow mobility is even seen among the terrestrial-adapted African apes and humans.

Forearms

The ulna is long in the brachiating apes, spider monkeys, and slender lorises. The ulna is more moderate in length among most other primates. Besides the length of the olecranon process, ulnae can be straight as in apes or slightly curved as in most other primates. In Old World monkeys, particularly terrestrial taxa, the ulna is often bent proximally relative to the shaft.

In primates like the terrestrial cercopithecines, the olecranon process is retroflexed, adding additional length to the triceps lever arm during extension. In apes, the ulnar trochlear notch faces anteroposteriorly compared to a more anteriorly orientation for most other primates (see fig. 8.15).

The radius is often moderate in length across primates and similar to humeral length in many taxa. The radius is also longer than the humerus in several varieties of primates, as noted above for the brachial index. Apes possess a long radial neck for pulling movements associated with climbing and suspension. Short radial necks can be found in cheirogaleids, tarsiers, the smaller galagos, and lorisids. This structural arrangement may help biceps brachii produce rapid flexion, a movement helpful with forelimb insectivory. In vertical clinging taxa the forelimbs are generally short and the radial head is strongly tilted. Radial heads are more circular among forelimb suspensory primates compared to a more oval anatomy observed across the majority of primates.

ODDITIES

Lorises have expanded their capitulum, reduced their trochlear regions, and generally possess a narrow distal humerus. Lorises have lost the zona conoidea as well and they possess a large coronoid fossa above

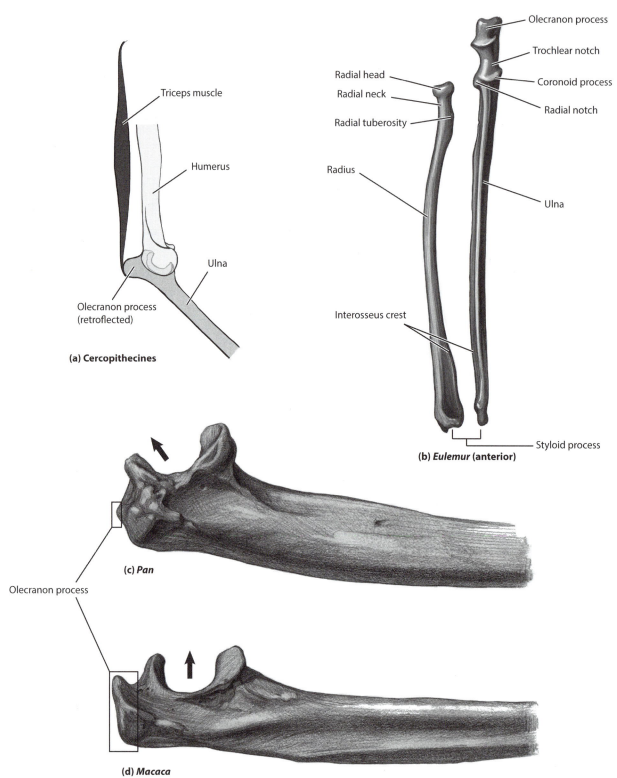

Triceps muscle

Humerus

Ulna

Olecranon process
(retroflected)

(a) Cercopithecines

Radial head

Radial neck

Radial tuberosity

Radius

Interosseus crest

Olecranon process

Trochlear notch

Coronoid process

Radial notch

Ulna

Styloid process

(b) *Eulemur* (anterior)

(c) *Pan*

Olecranon process

(d) *Macaca*

Figure 8.15 Proximal ulnar anatomy.

Primate elbow anatomy comes in two basic patterns with several anatomical varieties, depending on the lineage. These two basic patterns are also typical of fossil primates, and their anatomical similarities imply functional similarities as well. Illustrated are four elbow morphologies, two for fossil primates and two for living primates. On top, the elbow anatomy for the fossil catarrhine *Aegyptopithecus* from the Oligocene of Egypt is paired with the elbow of *Cebus*, a South American monkey. Below, the Eocene fossil North American omomyid elbow for *Shoshonius* is paired with the elbow of living *Tarsius* from Southeast Asia. Note the round capitulum, the gap, or zona conoidea, followed by the downward-angled trochlea for *Shoshonius* and *Tarsius*. This elbow joint suggests exceptional radial mobility for the hand and forearm for pronation and supination movements during climbing sequences. For the anthropoids, *Aegyptopithecus* and *Cebus*, note that the zona conoidea has been compressed and the capitulum is more flattened, being more cylindrical in shape. These shape changes are interpreted as implying increased use of arboreal quadrupedalism, relative to climbing, in anthropoids. Both examples of elbow anatomy show how we can use living primate anatomy to infer functional patterns among fossil primates.

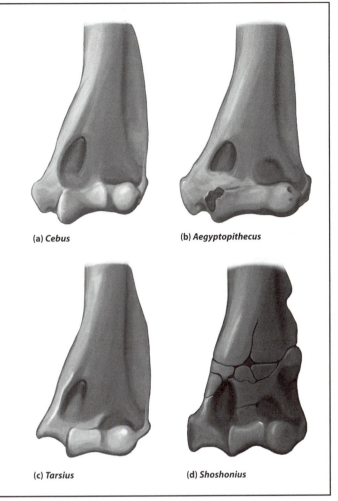

(a) *Cebus*   **(b)** *Aegyptopithecus*

(c) *Tarsius*   **(d)** *Shoshonius*

the capitulum for the radius. These features suggest a greater reliance on radial versus ulnar compression at the elbow.

As a group, cercopithecids are recognized for their reduction in joint rotations, focusing mainly on fore and aft movements. Their elbow, as well as their shoulder and hand, reflect these modifications.

In vertical clingers (e.g., indriids and tarsiers), the trochlea is taller than the capitulum and narrow in width in contrast to the elbow anatomy of *Daubentonia*, which is distinctive in possessing a large brachial flange on its distal humerus. In all other aspects the humerus of an aye-aye is very similar to that of other lemurs.

As noted above, the cercopithecids possess a narrow distal humerus with a capitulum and trochlea pushed together. The trochlea is highly modified in these taxa in that the medial trochlear rim, or keel, is extremely

long and pronounced. This feature, as well as the posterior trochlear's lateral keel, helps stabilize the humeroulnar joint in a flexed position. The olecranon fossa is quite deep, and the medial epicondyle is oriented posteriorly. All of these structural modifications make Old World monkey elbows quite distinct from those of other primates and appear to be related to a terrestrial quadrupedal ancestry. Many of these features are more pronounced among the terrestrial taxa alive today, especially in the long-limbed and fast patas monkey.

Hands

Primate hands represent one of the original and fundamental adaptations of primates (fig. 8.16). Hands are grasping structures and are generally palmigrade in their orientation to a support (see fig. 8.3). Some primates use digitigrade, knuckle-walking, suspensory grasps or fist-walking hand postures. Grasping hands must flex all five digits at the metacarpophalangeal and interphalangeal joints, while simultaneously opposing the first digit or thumb. Opposability, or the swing motion of the thumb, is accomplished at the first metacarpal-trapezium joint. Rotation of the hand, pronation or internal rotation, aligns the palm with the support, usually a branch. Hand rotation, either pronation or supination, an external rotation, occurs primarily at the elbow joint between the radial head and the capitulum with little mobility at the wrist joints. Wrist joints primarily function to flex and extend the hand.

Carpal, or wrist, bones in primates generally number nine (fig. 8.17; proximal row: scaphoid, centrale, lunate, triquetral, and pisiform; distal row: trapezium, trapezoid, capitate, and hamate). The centrale bone is fused to the scaphoid in African apes and humans; this fusion is related to weight transmission during knuckle walking in African apes and likely human ancestry as well. The centrale is also independently fused among indriids and *Lepilemur*, non-knuckle-walking taxa, however. Fused centrale taxa possess only eight carpal bones. The proximal capitate and hamate joint surfaces form a ball-like shape within the mid-carpal joint, facilitating mid-carpal rotations across primates.

Brachiating primates like spider monkeys, gibbons,

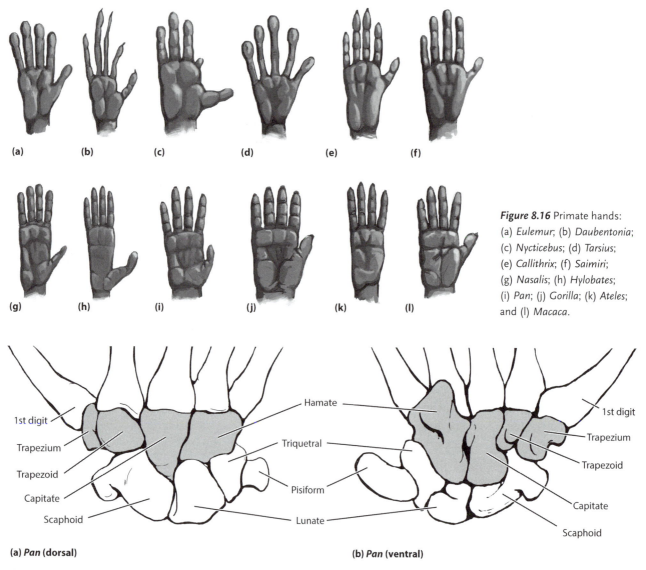

Figure 8.16 Primate hands:
(a) *Eulemur*; (b) *Daubentonia*;
(c) *Nycticebus*; (d) *Tarsius*;
(e) *Callithrix*; (f) *Saimiri*;
(g) *Nasalis*; (h) *Hylobates*;
(i) *Pan*; (j) *Gorilla*; (k) *Ateles*;
and (l) *Macaca*.

(a) *Pan* (dorsal)

(b) *Pan* (ventral)

Figure 8.17 Primate carpal anatomy (*Pan*).

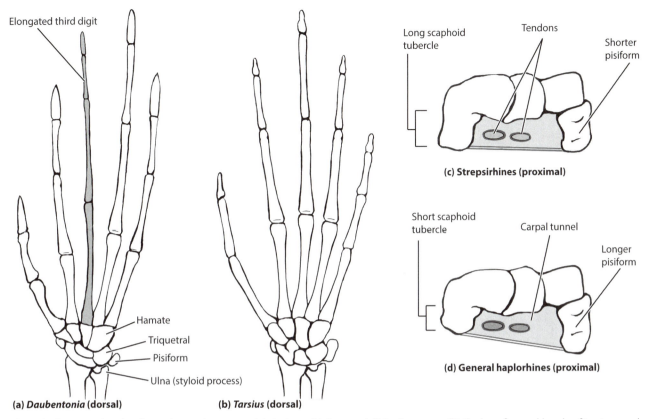

Figure 8.18 Primate hand and carpal tunnel anatomy: (a) the long third manual digit of aye-ayes; (b) the long-fingered hands of tarsiers; and (c) the deep carpal tunnel anatomy of strepsirhine primates relative to the carpal tunnel of haplorhine primates (d).

and orangutans possess a rotary mid-carpal region where the scaphoid can rotate around the medial capitate. Lorises also have several wrist features similar to that of apes. In contrast, the African apes and humans possess wrist features for weight bearing when using their hands for weight support during quadrupedalism.

Strepsirhine primates have shorter pisiforms and a more developed hamate hamulus than do haplorhine primates (fig. 8.18). Strepsirhines and tarsiers also possess a relatively longer scaphoid tubercle relative to that of anthropoid wrist anatomy. These wrist features document a deeper carpal tunnel for the extrinsic digital flexors in strepsirhines relative to haplorhines and are linked functionally to climbing and vertical support use in strepsirhines and to quadrupedalism among haplorhines. A relatively large hamate and centrale-hamate contact for strepsirhines has also been linked to ulnar deviation. Ulnar or laterally deviated hand positions in strepsirhines come from the mid-carpal joint, whereas in haplorhines this same

ulnar movement occurs at the metacarpophalangeal joint—another contrast between these two groups of primates.

Another distinctive wrist feature in primates is where the wrist bones contact the ulna. The styloid process at the distal end of the ulna is long and contacts the triquetrum and pisiform wrist bones in most primates. In apes, the styloid process is reduced in length and does not contact these two wrist bones. A pad, or fibrocartilaginous meniscus, is located between the ulnar styloid and the triquetrum in apes, preventing articulation between these bones. The reduction of the styloid process results in greater abduction at the wrist in apes.

Fingers

All primate fingers possess nails, although a few species possess claw-like or keeled nails. It is easy to tell a nailed primate from a clawed mammal osteologically since the distal phalanx of primates is flat and broad,

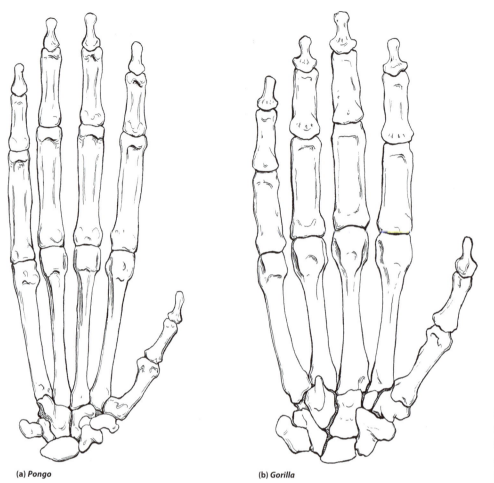

(a) *Pongo*   **(b)** *Gorilla*

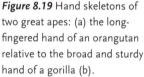

Figure 8.19 Hand skeletons of two great apes: (a) the long-fingered hand of an orangutan relative to the broad and sturdy hand of a gorilla (b).

relative to the narrow, tall, long, and curved bony phalangeal bone surface for a claw. Some primates have re-evolved "claws" or claw-like nails on their lateral digits of their hands (i.e., callitrichids, *Dauben-tonia, Phaner, Eulemur rubriventor*, and *Euoticus*). On the plantar surface of the distal phalanx for the thumb and big toe, there is a large depression for an expanded flexor tendon attachment area for the first digits of primates.

In terms of lateral finger lengths across primates, most scale finger length relative to mass, with smaller taxa possessing longer fingers relative to larger primates. This means that fingers (and toes) are relatively shorter as body size increases. This is not the case, however, for thumbs or big toes of primates, which retain similar lengths across increasing or decreasing size changes. Two primates, the tarsier and the aye-aye (see fig. 8.18), are characterized by very long fingers relative to those of other primates. Non-anthropoid

primates have long fourth digits (ectaxonic hands), while anthropoids possess long third digits (mesax-onic hands) (see fig. 8.16). The thumb in apes and Old World monkeys is completely opposable relative to the thumbs of other primates.

ODDITIES

Aye-ayes have the most unusual primate hand in that they possess a greatly elongated, slender, and highly mobile third digit with a claw at the end for piercing insect larvae (see fig. 8.18). Lorises also pos-sess odd hands with a greatly reduced second digit since they have shifted the position of their thumb to align itself across from the third digit of the hand. This increases the grasping span of loris hands pre-sumably for larger-diameter supports. Suspensory taxa, like orangutans or gibbons, have extraordinarily long and curved fingers (fig. 8.19), while other pri-mates (i.e., *Colobus, Ateles*, or *Pongo*; see fig. 8.16) have

Tarsius

Figure 8.20 The arched finger position of tarsier hands. Note the flat terminal finger position relative to the flexed arch position for the proximal and middle phalanges.

lost or reduced their first digit. The loss or reduction of the first digit makes the primate hand into a functional hook rather than a five-fingered grasping organ with an opposable thumb. Thumb loss or reduction therefore is more common across primates than is the loss of a grasping big toe, which occurs only in humans.

A few primates, like *Alouatta*, often grasp a support between their second and third digits (ectaxic) relative to their thumb, while indriids are known for divergent thumbs relative to those of other primates. Tarsiers are known for their wide and circular fingertips as well as a thumb that is described as non-opposable. Tarsier fingers, besides being long, are also odd in their ability to be arched (fig. 8.20). A few primates possess webbing between their digits (e.g., *Propithecus, Indri, Perodicticus, Arctocebus*, and *Hylobates syndactylus*).

Selected References

Aiello, L., and C. Dean. 1990. An Introduction to Human Evolutionary Anatomy. Academic Press, New York.

Ankel-Simons, F. 2000. Primate Anatomy—An Introduction. Academic Press, New York.

Ashton, E.H., and C.E. Oxnard. 1964. Functional adaptations of the primate shoulder girdle. Proceedings of the Zoological Society of London 142:49–66.

Beard, K.C., M. Dagosto, D.L. Gebo, and M. Godinot. 1988. Interrelationships among primate higher taxa. Nature 331:712–714.

Cant, J.G.H., D. Youlatos, and M.D. Rose. 2001. Suspensory locomotion of *Lagothrix lagothricha* and *Ateles belzebuth* in Yasuni National Park, Ecuador. Journal of Human Evolution 44:685–699.

Cartmill, M., 1972. Arboreal adaptations and the origin of the Order Primates; pp. 97–122 *in* R.H. Tuttle (ed.), The Functional and Evolutionary Biology of Primates. Aldine Press, Chicago.

———. 1985. Climbing; pp. 73–88 *in* M. Hildebrand, D.M. Bramble, K.F. Liem, and D.B. Wake (eds.). Functional Vertebrate Morphology. Belknap Press of Harvard University Press, Cambridge, Mass.

Cartmill, M., P. Lemelin, and D.O. Schmitt. 2002. Support polygons and symmetrical gaits in mammals. Zoological Journal of the Linnean Society 136:401–420.

Cartmill, M., and K. Milton. 1977. The lorisiform wrist joint and the evolution of "brachiating" adaptation in the Hominoidea. American Journal of Physical Anthropology 47:249–272.

Dagosto, M. 1983. Postcranium of *Adapis parisiensis* and *Leptadapis magnus* (Adapiformes, Primates). Folia Primatologica 41:49–101.

Dagosto, M., D.L. Gebo, and K.C. Beard. 1999. Revision of the Wind River faunas, Early Eocene of central Wyoming. Part 14. Postcranium of *Shoshonius cooperi* (Mammalia: Primates). Annals of Carnegie Museum 68(3):175–211.

Demes, B., S.G. Larson, J.T. Stern, W.L. Jungers, A.R. Biknevicius, and D. Schmitt. 1994. The kinetics of primate quadrupedalism: "hindlimb drive" reconsidered. Journal of Human Evolution 26:353–374.

Fleagle, J.G., and E.L. Simons. 1982. The humerus of *Aegyptopithecus zeuxis*, a primitive anthropoid. American Journal of Physical Anthropology 59:175–193.

———. 1995. Limb skeleton and locomotor adaptations of *Apidium phiomense*, an Oligocene anthropoid from Egypt. American Journal of Physical Anthropology 97:235–289.

Gebo, D.L. (ed.). 1989. Postcranial adaptation and evolution in Lorisidae. Primates 30(3):347–367.

———. 1993. Postcranial Adaptation in Nonhuman Primates. Northern Illinois University Press, DeKalb.

———. 1996. Climbing, brachiation, and terrestrial quadrupedalism: historical precursors of hominid bipedalism. American Journal of Physical Anthropology 101:55–92.

Gregory, W.K. 1920. On the structure and relations of *Notharctus*, an American Eocene primate. Memoirs of the American Museum of Natural History 3:49–243.

Gunther, M.M. 1985. Biomechanische vorassetzungen bein Absprung des Senegalagos. Zeitschrift für Morphologie und Anthropologie 75:287–306.

Gunther, M.M., H. Ishida, H. Kumakura, and H. Nakano. 1991. The jump as a fact mode of locomotion in arboreal and terrestrial biotopes. Zeitschrift für Morphologie und Anthropologie 78(3):341–372.

Hall-Craggs, E.C.B. 1965. An analysis of the jump of the lesser galago. Journal of Zoology (London) 147:20–29.

Hamrick, M.W. 1997. Functional osteology of the primate carpus with special reference to the Strepsirhini. American Journal of Physical Anthropology 104:105–116.

Hildebrand, M. 1967. Symmetrical gaits of primates. American Journal of Physical Anthropology 26:119–130.

Hunt, K.D., J.G.H. Cant, D.L. Gebo, M.D. Rose, S.E. Walker, and D. Youlatos. 1996. Standardized descriptions of primate locomotor and postural modes. Primates 37(4):363–387.

Jenkins, F.A. 1981. Wrist rotation in primates: a critical adaptation for brachiators. Symposium of the Zoological Society of London 48:429–451.

Jenkins, F.A., and J.G. Fleagle. 1975. Knuckle-walking and the functional anatomy of the wrists in living apes; pp. 213–227 in R. Tuttle (ed.), Primate Functional Morphology and Evolution. Mouton Publishers, Paris.

Jouffroy, F.K., M. Godinot, and Y. Nakano. 1993. Biometrical characteristics of primate hands; pp. 133–171 in H. Preuschoft and D.J. Chivers (eds.), Hands of Primates. Springer Verlag, New York.

Jouffroy, F.K., and J. Lessertisseur. 1979. Relationships between limb morphology and locomotor adaptations among prosimians: an osteometric study; pp. 143–182 in M.E. Morbeck, H. Preuschoft, and N. Gomberg (eds.). Environment, Behavior, and Morphology: Dynamic Interactions in Primates. Gustav Fischer, New York.

Jungers, W.L. 1985. Body size and scaling of limb proportions in primates; pp. 345–382 in W.L. Jungers (ed.), Size and Scaling. Plenum Press, New York.

Keith, A. 1923. Man's posture: its evolution and disorders. British Medical Journal 1:451–454, 545–548, 587–590, 624–626, 669–672.

Kimura, T., M. Okada, and H. Ishida. 1979. Kinesiological characteristics of primate walking: its significance in human walking; pp. 297–311 in M.E. Morbeck, H, Preuschoft, and N. Gomberg (eds.), Environment, Behavior and Morphology: Dynamic Interactions in Primates. Gustav Fischer, New York.

Larson, S.G. 1988. Subscapularis function in gibbons and chimpanzees: implications for interpretation of humeral head torsion in hominoids. American Journal of Physical Anthropology 76:449–462.

———. 1998. Unique aspects of quadrupedal locomotion in nonhuman primates; pp. 157–173 in E. Strasser, J.G. Fleagle, A.L. Rosenberger, and H.M. McHenry (eds.), Primate Locomotion: Recent Advances. Plenum Press, New York.

Larson, S.G., D. Schmitt, P. Lemelin, and M. Hamrick. 2000. Uniqueness of primate forelimb posture during quadrupedal locomotion. American Journal of Physical Anthropology 112:87–101.

———. 2001. Limb excursion during quadrupedal walking: how do primates compare to other mammals? Journal of Zoology (London) 255:353–365.

Lemelin, P., and W.L. Jungers. Body size and scaling of the hands and feet of prosimian primates. American Journal of Physical Anthropology 133:828–840.

Lemelin, P., and D. Schmitt. 1998. The relation between hand morphology and quadrupedalism in primates. American Journal of Physical Anthropology 105:185–197.

Lewis, O.J. 1969. The hominoid wrist joint. American Journal of Physical Anthropology 30:251–268.

———. 1989. Functional Morphology of the Evolving Hand and Foot. Oxford Science Publications, Oxford, UK.

Morbeck, M.E., H. Presuschoft, and N. Gomberg (eds.). 1979. Environment, Behavior, and Morphology: Dynamic Interactions in Primates. Gustav Fischer, New York.

Napier, J.R. 1967. Evolutionary aspects of primate locomotion. American Journal of Physical Anthropology 27:333–342.

———. 1980. Hands. Pantheon Books, New York.

Napier, J.R., and A.C. Walker. 1967. Vertical clinging and leaping—a newly recognized category of locomotor behavior of primates. Folia Primatologica 6:204–219.

Reynolds, T.R. 1985. Stresses on the limbs of quadrupedal primates. American Journal of Physical Anthropology 67:351–362.

———. 1987. Stride length and its determinants in humans, early hominids, primates and mammals. American Journal of Physical Anthropology 72:101–116.

Rollinson, J., and R.D. Martin. 1981. Comparative aspects of primate locomotion, with special reference to arboreal cercopithecines. Symposium of the Zoological Society of London 48:377–427.

Rose, M.D. 1983. Miocene hominoid postcranial morphology: monkey-like, ape-like, neither, or both?; pp. 405–433 in R.L. Ciochon and R.S. Corruccini (eds.), New Interpretations of Ape and Human Ancestry. Plenum Press, New York.

———. 1988. Another look at the anthropoid elbow. Journal of Human Evolution 17:193–224.

Ruff, C.B., and J.A. Runestad. 1992. Primate limb bone structural adaptations. Annual Review of Anthropology 21:407–433.

Sarmiento, E.E. 1988. Anatomy of the hominoid wrist joint: its evolutionary and functional implications. International Journal of Primatology 9(4):281–345.

Schmitt, D. 1995. A kinematic and kinetic analysis of forelimb use during arboreal and terrestrial quadrupedalism in Old World monkeys. Ph.D. Dissertation, State University of New York at Stony Brook.

Schmitt, D., and P. Lemelin. 2002. Origins of primate locomotion: gait mechanics of the woolly opossum. American Journal of Physical Anthropology 118:231–238.

Schultz, A.H. 1969. The Life of Primates. Weidenfeld and Nicolson, London.

Swartz, S.M. 1993. Biomechanics of Primate Limbs; pp. 5–42 *in* D.L. Gebo (ed.), Postcranial Adaptation in Nonhuman Primates. Northern Illinois University Press, DeKalb.

Szalay, F.S., and M. Dagosto. 1980. Locomotor adaptations as reflected on the humerus of Paleogene primates. Folia Primatologica 34:1–45.

Terranova, C.I. 1995. Functional morphology of leaping behaviors in galagids: associations between landing limb use and diaphyseal geometry; pp. 473–493 *in* L. Alterman, G.A. Doyle, and M.K. Izard (eds.), Creatures of the Dark—The Nocturnal Prosimians. Plenum Press, New York.

Tuttle, R.H. 1967. Knuckle-walking and the evolution of hominoid hands. American Journal of Physical Anthropology 26:171–206.

————. 1969. Quantitative and functional studies on the hands of the Anthropoidea. Journal of Morphology 128:309–363.

————. 1970. Postural, propulsive, and prehensile capabilities in the cheiridia of chimpanzees and other great apes; pp. 167–253 *in* G.H. Bourne (ed.), The Chimpanzee, Vol. 2. Karger, Basel.

Vilensky, J.A. 1989. Primate quadrupedalism: how and why does it differ from that of typical quadrupeds? Brain, Behavior and Evolution 34:357–364.

Washburn, S.L. 1968. The Study of Human Evolution (Congdon Lectures). University of Oregon Books, Eugene.

Whitehead, P. 1993. Aspects of the wrist and hand; pp. 96–120 *in* D.L. Gebo (ed.), Postcranial Adaptation in Nonhuman Primates. Northern Illinois University Press, DeKalb.

9 Hindlimb

HINDLIMB

Hindlimbs are the key to primate locomotion. As noted in chapter 8, primates are hindlimb driven and their legs can be quite elongated, especially in the specialized leaping primates like tarsiers, galagos, and indriids. Although primates use their hands for foraging and manipulating objects, few use their arms exclusively to propel their bodies forward (with the exception of the brachiating primates). Legs are the key propulsive limb element and this is especially true for humans. Another general primate hindlimb feature is the abducted position of the lower limbs (fig. 9.1), a position that moves the knees away from the midline of the body in both horizontal and vertical body positions. Likewise, the asymmetrical shape of hindlimb joints is a notable structural component of primate limbs and a key adaptation for arboreality.

Primate legs can vary in length, being long (e.g., tarsiers and galagos), average in length (e.g., squirrel and cebus monkeys), or relatively short (e.g., cheirogaleids or orangutans). Long-legged primates are good, if not spectacular, leapers. Primates with average leg lengths tend toward quadrupedal movements, while short legs are often associated with climbing primates, although there are exceptions (e.g., the short-legged cheirogaleids). Long legs are also found in terrestrial cursors like patas monkeys, whose long legs increase stride length. Similarly, slender lorises also possess relatively elongated hindlimbs, like their forelimbs, but they are neither leapers nor cursors; they exercise

unusual but highly flexible movement patterns. In contrast to slender lorises, African apes are unusual in that they possess shorter legs relative to their long arms, creating a back posture of 45 degrees when knuckle walking (fig. 9.2), compared to the more common horizontal orientation of most other primate backs.

Cheirogaleids, some of the smallest living primates, have surprisingly short limbs with several species being capable of frequent leaping, as observed in their long-legged cousins the lemurs. In fact, cheirogaleids seem to spread their movements out along the major locomotor categories rather than specializing in any single movement type. Cheirogaleids are able to move quadrupedally and utilize climbing and leaping like lemurs, but they are also capable of using quadrupedal suspensory movements frequently in comparison to the larger-sized lemurs (fig. 9.3). Cheirogaleid grasping strength is such that they often hold on to and move along the sides or the underside of branches without falling, as well as being able to cantilever (see fig. 8.7), a horizontal body position requiring considerable foot and leg strength. The short, bent limbs of cheirogaleids place their bodies close to a support and this body position allows for better balance over relatively small support diameters. It is not certain why short limbs would be an adaptive response to move along the sides of or below a branch in comparison to other primate movement patterns.

In contrast to the short-legged cheirogaleids, vertical clinging and leaping primates have long legs. Vertical clingers and leapers come in three varieties:

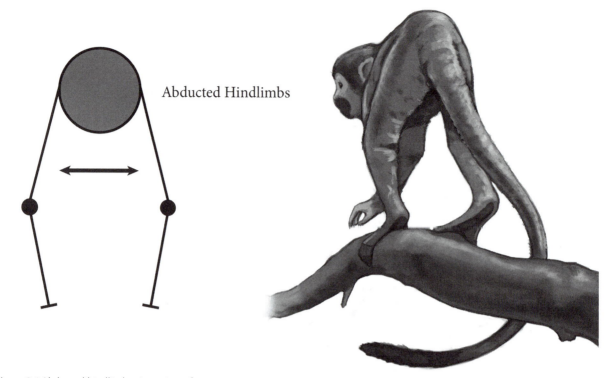

Abducted Hindlimbs

Figure 9.1 Abducted hindlimb orientation of primates.

long legs and long feet (galagos and tarsiers; fig. 9.4), long legs and short feet (indriids and *Lepilemur*), and average leg length with clawed feet (callitrichines). The locomotor pattern, vertical clinging and leaping, has clearly evolved independently in separate primate clades with differing anatomical characteristics. Only the callitrichines (figs. 9.4–9.6) use this mode on large-diameter tree trunks. Most, active vertical clingers and leapers, prefer to move below the canopy. For example, tarsiers prefer tiny vertical saplings (< 3 m in height) and move and cling to them close to the ground. Primates like *Hapalemur* and *Callimico* vertical-cling and -leap through bamboo, while Indriids and *Lepilemur* use this same movement pattern higher in the canopy. The New World monkey, *Pithecia*, is also known to vertical-cling and -leap through the lower canopy frequently. In fact, vertical clinging and leaping has been viewed as the initial locomotor pattern for all primates.

THE PELVIS AND HIP JOINT

As noted in chapter 7, the primate pelvis is made up of three bones: the ilium, the ischium, and the pubis. All meet and fuse into a single bone at the acetabulum, the actual hip joint. The pelvis of primates comes in two standard anatomical varieties: a rod-like ilium or an expanded, rectangular-shaped ilium (fig. 9.7). The rod-like ilium is found in many of the non-anthropoid primates, but not all. Galagos, lorises, cheirogaleids, *Lepilemur*, *Daubentonia*, and tarsiers have this type of pelvic shape. In these taxa the ilium is long, the iliac crest is relatively narrow, and the gluteal surface is slightly concave facing laterally. Tarsiers, indriids, and galagos have the longest ilia among prosimian primates. In contrast, lemurs and indriids have widened the cranial aspect of the ilium so that this bone looks similar to that of an anthropoid pelvis, with a widened, board-shaped iliac blade (fig. 9.7). In lemurs and indriids, the iliac blade is wide cranially with extended iliac projections in contrast to those of anthropoids, which begin iliac widening at the region just above the acetabulum continuing cranially upward. In cercopithecines the ilium is especially curved (concave), whereas apes have widely expanded and flattened the iliac surface mediolaterally (fig. 9.8). A mediolaterally expanded and flat ilium moves the gluteal muscle's angle of contraction laterally outward, away from the

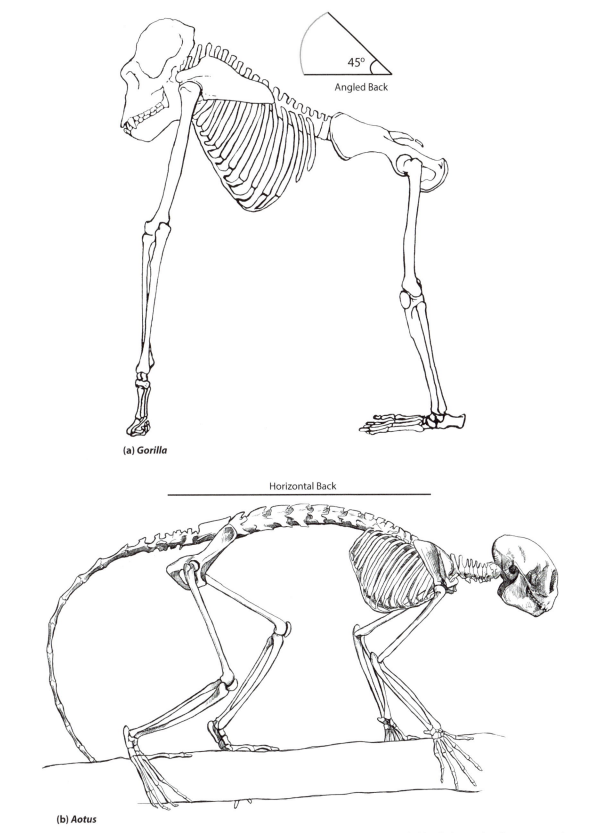

45°

Angled Back

(a) *Gorilla*

Horizontal Back

(b) *Aotus*

Figure 9.2 Back orientation in primates. Knuckle-walking primates (a, *Gorilla*) possess an angled back due to their long arms, relative to their legs, in comparison to the horizontal backs of most primates with more similar limb lengths (b, *Aotus*).

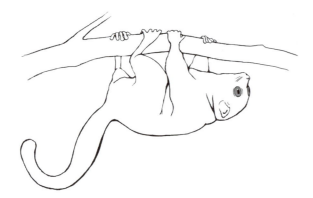

Figure 9.3 Suspensory quadrupedalism among cheirogaleids (*Microcebus*)

spine. In contrast, a curved or concave-shaped ilium packs muscles into a semi-cylindrical surface (e.g., cercopithecids; fig. 9.8).

Among anthropoids, including the small callitrichids, the wide ilium, or gluteal plane, increases the area of origin and mass for the gluteus medius, the largest gluteal muscle among non-human primates. Jack Stern from State University of New York at Stony Brook has proposed that this muscle is a medial rotator of the thigh (or a lateral rotator of the pelvis) and that the gluteus medius acts as a hip stabilizer. Hip stability may be an important functional consideration when thinking about limb mechanics among anthro-

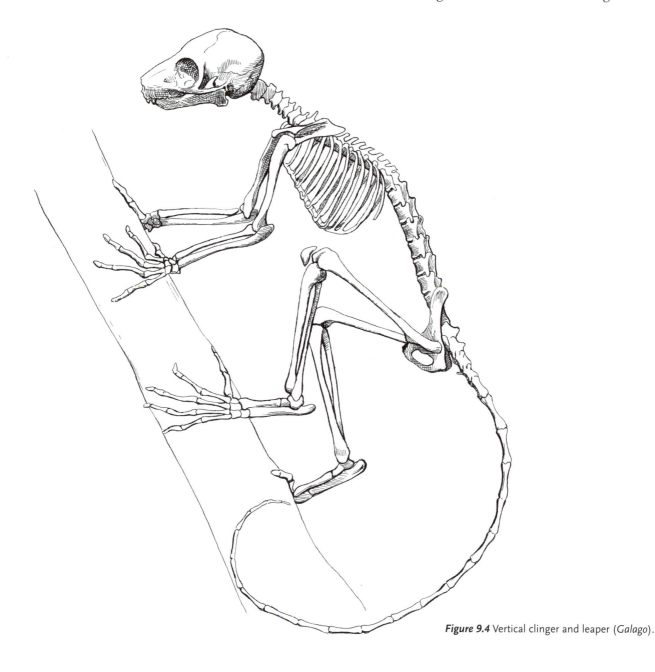

Figure 9.4 Vertical clinger and leaper (*Galago*).

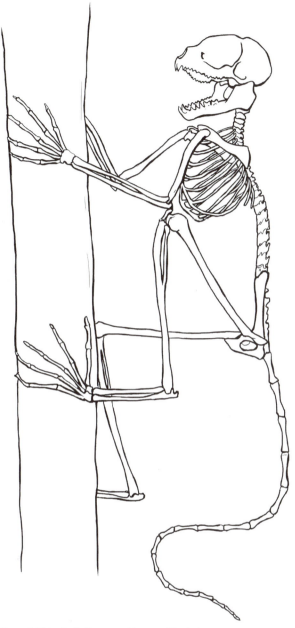

Figure 9.5 Vertical clinger and leaper (*Tarsius*).

Figure 9.6 Vertical clinger and leaper (*Leontopithecus*).

poid arboreal quadrupedalists. As one might surmise, the length of the primate ilia affects the mechanics of the hip joint, while the width of the ilia affects total muscle mass. A tall and wide ilium, found among anthropoids, allows a long moment arm to be associated with a larger muscle mass, particularly for the gluteus medius. Leaping primates are characterized by a relatively longer ilium that increases the length of the tensor fascia lata moment arm, a muscle involved in flexing the femur at the hip joint during the recovery stroke.

Anthropoid ilia also demonstrate a wide iliac plane and a prominent ridge, the margo acetabuli (fig. 9.8), which separates the iliac and gluteal planes relative to prosimian primates. In a somewhat similar manner, lemurs and indriids have proximally splayed their iliac crest regions, thereby increasing the surface area for the gluteus medius and iliacus muscles.

The ischium part of the pelvis shows elongation or flattening across the living primates. Old World monkeys and gibbons show ischial callosities, a flattened ischium where the soft tissues for their leathery sitting pads attach (fig. 9.9). Vertical clinging and leaping primates possess short ischia, which are bent dorsally in galagos, indriids, and tarsiers. Tarsiers possess the longest ilium and the shortest ischium compared to those of the other strepsirhine leapers (fig. 9.9).

The length of the ischium approximates the moment arm for the hamstring muscles (the semitendinosus and the semimembranosus). Hip extension is an important propulsive force in leaping primates.

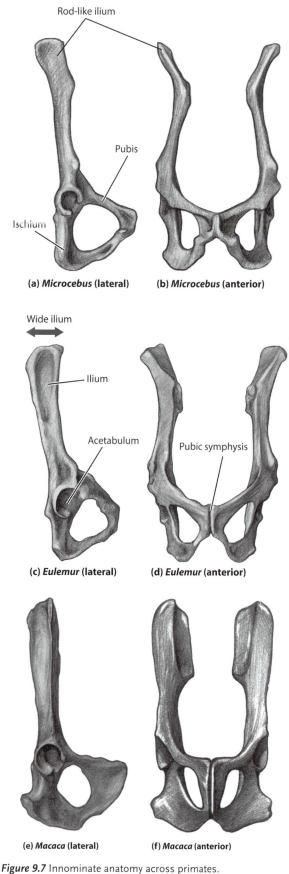

Rod-like ilium

Pubis

Ischium

(a) *Microcebus* (lateral)   **(b) *Microcebus* (anterior)**

Wide ilium

Ilium

Acetabulum

Pubic symphysis

(c) *Eulemur* (lateral)   **(d) *Eulemur* (anterior)**

(e) *Macaca* (lateral)   **(f) *Macaca* (anterior)**

Figure 9.7 Innominate anatomy across primates.

Short ischia are found among primates that require fast accelerations during hip extension (i.e., leapers), although the slow-climbing lorises also possess this anatomical feature, as do humans. In contrast, a long ischium results in a long lever arm, increasing hamstring musculature force during hip extension, and is generally in line with the gluteal plane of the ilium for quadrupedal leapers.

The pubis bone is long in lorises, followed by tarsiers and cheirogaleids among the prosimians. Hip adductor musculature attaches to this bone and its length implies differences in adductor muscle function across primates. Unfortunately, I know of no functional studies that have correlated pubic length with a specific locomotor adaptation.

Two other structural features of the ilium are important: the tubercle for the attachment of the rectus femoris muscle and the acetabulum, the actual hip joint and the point of attachment for the leg. The rectus femoris tubercle (fig. 9.10), a bony bump that lies above the acetabulum on the ilium, is large and prominent among indriids and *Lepilemur*, prominent in lemurs, but smaller in cheirogaleids, galagos, and *Daubentonia*. This tubercle is smaller still in the leaping tarsiers and tiniest among anthropoids and lorises. This tubercle is the attachment site for the rectus femoris muscle, one of the muscles of the thigh involved in leg extension and hip flexion. Among the larger vertical clingers and leapers, the indriids and *Lepilemur*, is a prominent tubercle associated with a large muscle. This makes sense considering the propulsive leg extension needed to leap. Other leaping primates like lemurs or galagos also have good-sized tubercles. Tubercle prominence, or the lack thereof, also correlates with reduced or less forceful leaping among anthropoids and the non-leaping lorises. In contrast, the small-bodied tarsiers are frequent vertical clingers and leapers but possess a comparatively smaller tubercle than indriids or galagos. This suggests that tarsiers must be compensating for their lack of or reduced rectus femoris muscle force with an alternative thigh muscle during leaping.

The acetabulum, or hip joint, comes in two varieties in primates. The first type is common and is a dorsally buttressed joint for more quadrupedally oriented primates. Here the dorsal part of the acetabular facet is larger than the ventral part within the acetabulum.

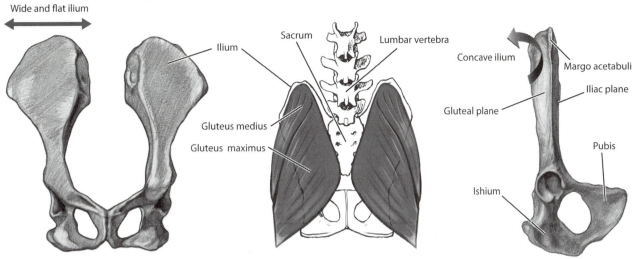

(a) *Hylobates* **(anterior)**

(b) *Hylobates* **(posterior)**

(c) *Macaca* **(lateral)**

Figure 9.8 Ilium anatomy.

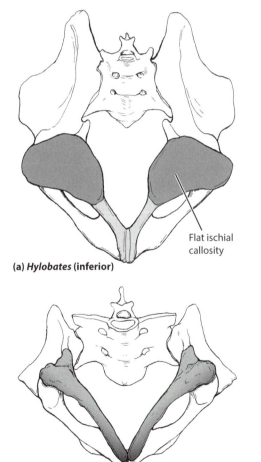

(a) *Hylobates* **(inferior)**

(b) *Eulemur* **(inferior)**

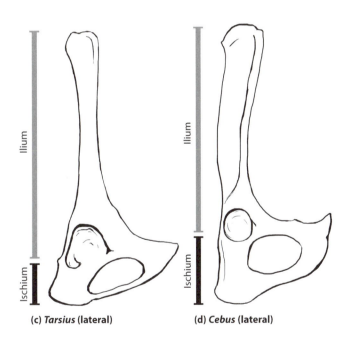

(c) *Tarsius* **(lateral)**

(d) *Cebus* **(lateral)**

Figure 9.9 Primate ischium. Note the expanded ischial callosities of gibbons (a) and Old World monkeys compared to those of most other primates (b, *Eulemur*). (c) and (d) contrast ilium to ischium lengths.

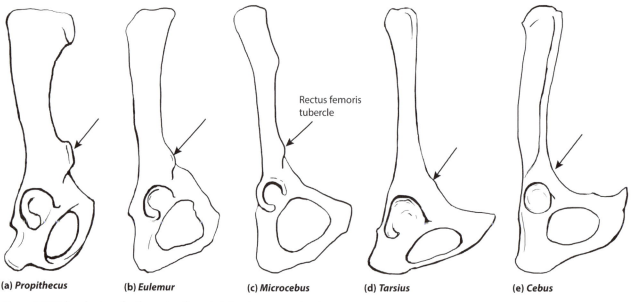

Rectus femoris
tubercle

(a) *Propithecus*   **(b)** *Eulemur*   **(c)** *Microcebus*   **(d)** *Tarsius*   **(e)** *Cebus*

Figure 9.10 Tubercle sizes for the rectus femoris tubercle across primates.

The second type of a primate hip joint is a ventrally buttressed joint with a large ventral facet, being observed in vertical clinging and leaping primates with well-abducted thighs and knees that are oriented cranially.

Oddities

The human pelvis is greatly shortened craniocaudally relative to that of apes, being essentially bowl-shaped. This and many other structural modifications in the human pelvis are adaptations for bipedalism (see chapter 10).

FEMUR

The femur, or thigh bone, articulates with the pelvis at the acetabulum, or hip joint. The hip joint is the starting point of leg mobility relative to the vertebral column. Primates that move in different ways have modified their femoral head morphology accordingly. The femoral head is generally round in most primates, giving them good overall mobility in several planes of motion (fig. 9.11). The femoral head often has a joint surface that extends toward the back side of the greater trochanter, allowing greater hip abduction. Femoral head morphology is heavily modified in galagos and tarsiers, which possess a more flattened

and cylindrical-shaped femoral head designed primarily for fore and aft mobility during specialized leaping (fig. 9.11). Lemurs and indriids, also frequent leaping primates, however, possess round femoral heads. Apes have enlarged, ball-like femoral heads for good overall hip mobility, allowing their legs to be placed in many different orientations.

Lorises have an interesting femoral head structure. Their femoral head is fairly cylindrical like that of galagos, although more rounded obliquely. Yet these primates do not leap as do galagos. Like galagos, the medial surface of the femoral head, where the fovea capitus resides, is flattened in lorises and galagos rather than being round or ball-like as in most other primates. Since lorises are known to move in a climbing, clambering, or serpentine quadrupedal fashion, requiring excellent leg and hip mobility, it is surprising that the loris hip joint has been modified like that of galagos, in ways that would limit overall joint movements. Since the current morphological situation seems inconsistent with the way lorises move today, it is reasonable to assume that lorises inherited this femoral head anatomy from a more leaping-oriented galago-like ancestor.

The femoral neck can be sorted into several morphological varieties across primates. Most primates have a femoral neck that is shaped like a cylindrical stalk, while leaping primates generally have a short

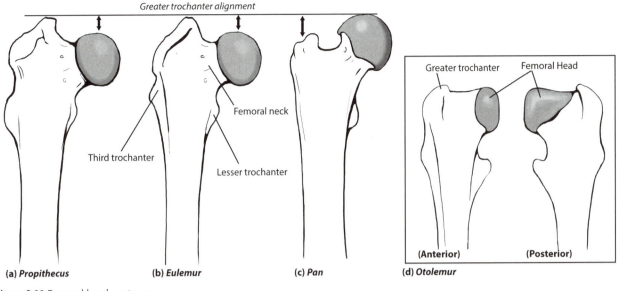

Figure 9.11 Femoral head anatomy.

and robust femoral neck (fig. 9.11). Climbing-oriented primates, like orangutans, have a long and highly angled femoral neck (fig. 9.12), while leaping and running primates tend to possess a horizontally oriented femoral neck. The angle of the femoral neck relative to the shaft affects the degree to which the hindlimb is abducted relative to the torso. The higher angles of most primates yield the characteristic slightly abducted hindlimb posture. In contrast, terrestrial cercopithecines have a more perpendicular orientation of the femoral head and neck relative to the shaft and greater trochanter (fig. 9.12), and thus a more adducted leg posture relative to other primates.

The femoral shaft can be elongated, round, or flattened across primates. The femoral shaft is quite long in a variety of primates, especially leaping primates like the indriids, and it is relatively short in the great apes, especially the orangutan. There is a tendency to flatten the femoral shaft mediolaterally in primates that climb or ones that suspend themselves from their legs. These primates also possess thicker cortical bone mediolaterally within the cross section of the femoral shaft. In contrast, an anteroposteriorly tall femoral shaft is found among leaping primates and other fast-moving quadrupedal taxa. These two different interior shaft designs counter forces being applied to the femur. Thicker cortical bone in flattened femoral shafts counters mediolateral, or side-to-side, forces while anteroposteriorly tall shafts counter forces being applied from a front to back direction.

The femoral head of primates is distinctive among galagos and tarsiers in terms of its semi-cylindrical head shape. As unusual as this shape may be, several fossil primates among the omomyid lineage also possess this distinctive femoral head morphology. Illustrated is the femoral head shape of *Hemiacodon gracilis*, an Eocene fossil primate from North America, next to that of a tarsier (see fig. 9.11 for a galago). The galago possesses the most extreme femoral head shape, followed in turn by *Hemiacodon* and *Tarsius*. The compression of the femoral head into a cylindrical shape relative to that of a round ball implies greater use of a fore and aft movement plane relative to a more circular hip movement pattern for a circular joint. We infer that the fossil *Hemiacodon* used its hip in a similar way and was clearly adapted to be a frequent leaping primate.

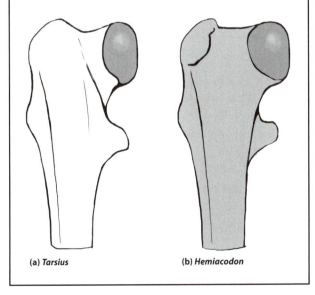

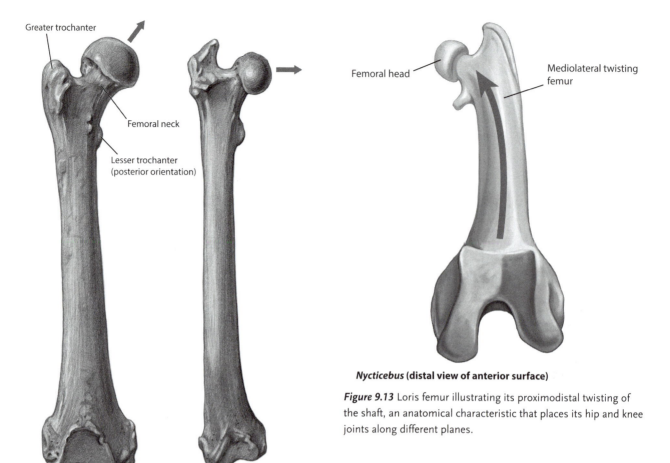

Figure 9.12 Primate femora.

(a) *Pongo*

(b) *Mandrillus*

Greater trochanter

Femoral neck

Lesser trochanter
(posterior orientation)

Femoral head

Mediolateral twisting
femur

Nycticebus **(distal view of anterior surface)**

Figure 9.13 Loris femur illustrating its proximodistal twisting of the shaft, an anatomical characteristic that places its hip and knee joints along different planes.

Compared to the other primates, lorises are again unusual in that their entire femoral shaft is twisted mediolaterally (fig. 9.13). This torsion places the femoral head in a different plane of orientation, being oriented medially toward the pelvis and spine relative to that of the knee joint, which places the lower leg in a laterally rotated alignment relative to the loris hip joint. The midline of the patellar facet lines up with the small third trochanter and the lateral edge of the greater trochanter in lorises relative to the usual primate alignment with the medial edge of the greater trochanter. One can only assume that this twisting is related to the many acrobatic positions utilized by lorises as they move through the canopy.

The primate femur may possess up to three trochanters, the greater, the lesser, and the third trochanter (see fig. 9.11), all of which anchor leg and hip musculature. The greater trochanter of primates can extend above, be in line with, or lie below the femoral

head. The greater trochanter is placed well above the femoral head in cercopithecines and in primates like lemurs, indriids, *Lepilemur*, and tarsiers (see fig. 9.11). The position of these two structures is more equal in galagos and *Daubentonia*, while in cheirogaleids the greater trochanter and the femoral head can be either equal in height or with the femoral head slightly above the greater trochanter. In contrast, in apes, especially orangutans, the femoral head is positioned well above the greater trochanter (fig. 9.12), as it is in lorises and platyrrhines.

The greater trochanter often overhangs the femoral shaft anteriorly and this proximal region of the femoral shaft is craniocaudally buttressed among many leaping primates. The greater trochanter is an important insertion point for the gluteus medius and the gluteus minimus, two of the largest hip muscles. Both act as thigh extensors in non-human primates. The gluteus maximus has a long line of insertion beginning along the posterolateral aspect of the greater trochanter and running distally down along the posterior femur. The gluteus maximus is often subdivided in primates into a thin and smaller part called the gluteus maximus proprius and a larger more caudal part called the ischiofemoralis that also acts as an extensor of the hip joint. Smaller hip muscles like the piriformis, the obturator internus, and the gemellus superior and inferior attach to the inferomedial surface of the greater trochanter as well and are involved in lateral rotation or abduction movements of the thigh.

The lesser femoral trochanter in primates is a prominent bony projection with different angles of orientation. The lesser trochanter is the main attachment site for the iliopsoas muscle, an important hip flexor and a key muscle during climbing. The lesser trochanter is often oriented medially among strepsirhines and tarsiers, whereas in anthropoids the lesser trochanter is generally smaller and obliquely oriented in a more posterior direction (see figs. 9.11 and 9.12). The lesser trochanter is posteriorly oriented among the cercopithecids. The lesser trochanter is often a prominent bony tubercle, but it can be flattened (e.g., lorises) or small as in several anthropoids.

Third trochanters (see fig. 9.11) are found in strepsirhines, but are much reduced in lorises and tarsiers. A small ridge can occasionally be found among platyrrhines, but this trochanter is absent in cercopithecids and apes. The gluteus maximus muscle attaches to the third trochanter. The third trochanter is proximally located in the leaping strepsirhine primates, especially the vertical clingers and leapers, and tarsiers, which are also vertical clinging and leaping primates. This trochanter's location is more distally located among lorises and platyrrhines, when present. A more proximal position for the third trochanter places the insertion site for gluteus maximus closer to the hip joint, thereby maximizing speed for hip and thigh extension, an important factor in achieving better takeoff velocity in leaping primates.

Oddities

The femoral head of galagos is a highly modified anatomical structure relative to that of other primates (see fig. 9.11). This cylindrical shape is clearly an adaptation for fore and aft movements at the hip joint during leaping. The long femoral neck and large, ball-like femoral head of orangutans, with its subsequent loss of the fovea capitus, makes this joint surface unusual as well among living primates. This type of

joint anatomy is clearly an adaptation for increased joint mobility at the hip for the extremely acrobatic leg positions utilized by orangutans.

KNEES

The knee joint functions primarily to bend (flex) or straighten (extend) the leg. Some twisting and sliding can also occur at this joint, but these movements are limited due to the muscular attachments and ligaments in this part of the leg. The distal femur has three curved joint surfaces: the medial and lateral condyles, and a joint surface for the patella (fig. 9.14). The lateral femoral condyle is round and fairly straight anteroposteriorly relative to the shaft. In contrast, the medial condyle is more S-curved. The patellar joint surface is broad and flat in lorises (fig. 9.14), apes, and platyrrhines, but is narrower in most other primates. The shape of the patellar joint surface is also reflected in the flatness, height, and length of primate patellae. In strepsirhines and tarsiers, the lateral rim of the

The knee joint of primates is often quite tall, exhibiting a high lateral patellar rim among frequent leaping primates. This type of a knee joint is common among living strepsirhines and for *Tarsius*. It is often observed among Eocene fossil primates as well. Illustrated below are knee joints for the North American adapiform fossil primate *Cantius trigonodus* and for the North American omomyid fossil primate *Hemiacodon gracilis*. Note that both fossil primates, which are from different evolutionary lineages, display a similar knee joint anatomy with all non-lorisine strepsirhines (see fig. 9.14). This tall knee pattern implies frequent leaping adaptations across all lineages and represents the primitive condition for all primates, given that this knee shape is found in the most primitive members of each suborder, Strepsirhini and Haplorhini.

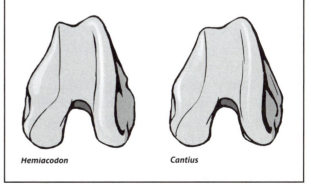

Hemiacodon   *Cantius*

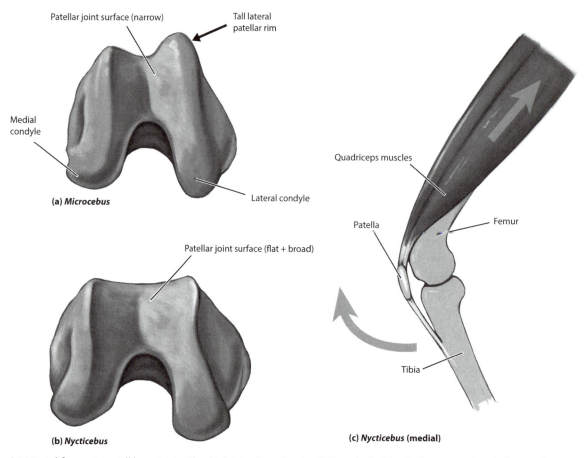

Patellar joint surface (narrow)

Tall lateral
patellar rim

Medial
condyle

Lateral condyle

(a) *Microcebus*

Patellar joint surface (flat + broad)

(b) *Nycticebus*

Quadriceps muscles

Patella

Femur

Tibia

(c) *Nycticebus* (medial)

Figure 9.14 Distal femur: (a) a tall knee joint with a high lateral patellar rim (*Microcebus*); (b) a flat knee joint (*Nycticebus*); and (c) quadriceps muscle contraction for the elevation of the lower leg (extension).

patellar facet is elevated (fig. 9.14). This elevation helps to prevent the patella from dislocating as it is pulled upward by the quadriceps muscles. This group of four large muscles, all of which lie on the anterior and mediolateral aspects of the femoral shaft, function to extend the leg, an especially important action for leaping primates.

Inferior to the distal femur is the joint surface for the proximal tibia. The proximal tibia in primates is a wide and flat surface with two anterior surfaces, the medial and lateral condyles, and an eminence in between (fig. 9.15). The lateral condylar facet of the tibia is generally convex in strepsirhines and tarsiers, and it extends farther posteriorly relative to anthropoids. This facet is most often concave in anthropoids. On the opposite side, the medial tibial condylar facet is convex and relatively oval or kidney-bean in shape. In terms of overall facet size, primates are highly variable. Lorises, *Microcebus*, and *Tarsius* have larger lateral facets in direct contrast to taxa such as *Colobus,*

Cercocebus, Cebus, or *Eulemur,* primates with larger medial facets.

Between the two proximal tibial facets lies the intercondylar eminence, a bony structure normally made up of two spines, a medial and a lateral spine, in most mammals. In strepsirhine primates, however, there is only a single large spine (fig. 9.15). In tarsiers and callitrichines, the medial spine is greatly reduced. Most platyrrhines and all catarrhines have the normal mammalian two-spine eminence with a groove in between. The single spine condition among strepsirhine primates reflects an increased use of lateral knee rotation around the intercondylar eminence when these primates place their leg on a vertical support. Tarsiers and callitrichines, which both possess a reduced medial spine, are also frequent users of vertical supports and appear to be mimicking the single spine functional condition among strepsirhines.

The knee joint has no sockets or interconnecting bony joint surfaces to keep these large bones in articulation. A series of ligaments holds the femur and tibia together. Medial and lateral ligaments, called the collateral ligaments, and the anterior and posterior cruciate ligaments, ligaments that lie in between the intercondylar notch of the distal femur and the tibial plateau, are the primary means of attachment for these two rather large bones of the leg (fig. 9.16). Without these ligaments or braces, the femur would simply slide off the tibial plateau.

Primate knees come in three anatomical shapes: (1) tall: distal femora with large, pronounced lateral patellar rims (non-anthropoid leapers); (2) average or normal height: distal femora with about equal condylar height to width dimensions (most living anthropoids); and (3) flattened: wide distal femora with wide patellar facets (lorises and living apes). Tall knees, or

knees with increased condylar height, provide greater mechanical advantage for the quadriceps muscle in knee extension. In contrast, low or flat knees are associated with frequent climbing primates that utilize a high variety of leg positions when grasping and pushing off from odd-angled supports. Average height knees are in between in their functional demands and are associated generally with quadrupedally oriented primates. As mentioned above (see fig. 9.13), the knees of lorises are angled laterally outward relative to their hip joints, allowing even further lateral rotations for the lower leg and foot.

A backward-tilted tibial plateau is present in all primate lineages, but this angle is quite pronounced, being retroflexed, among the frequent leaping primates like galagos and tarsiers (fig. 9.17). This tilt allows the

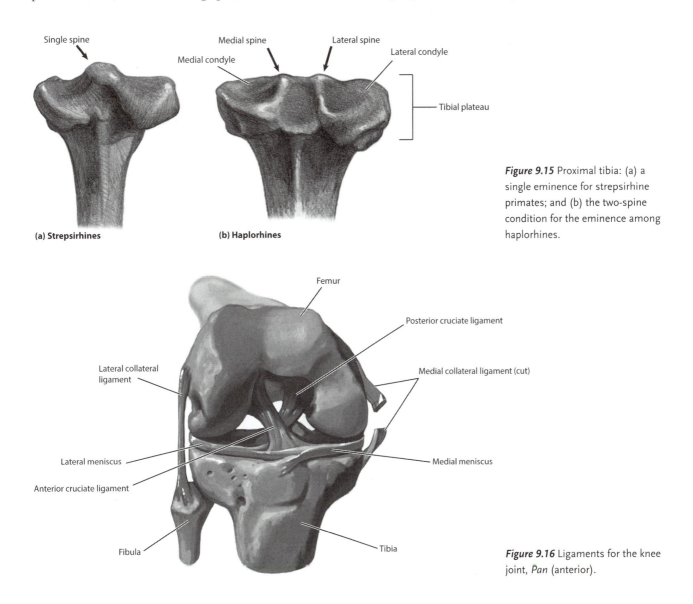

(a) Strepsirhines

(b) Haplorhines

Figure 9.15 Proximal tibia: (a) a single eminence for strepsirhine primates; and (b) the two-spine condition for the eminence among haplorhines.

Figure 9.16 Ligaments for the knee joint, *Pan* (anterior).

femoral condyles to rock backward for a larger angle of displacement during knee flexion.

The tibial shaft in primates is bowed proximally but is otherwise fairly straight in lateral view. Apes possess a greater degree of bowing, while Old World monkeys possess relatively straighter tibial shafts. Perhaps the most interesting tibial shaft shape in primates is that of the lorises. The tibial shaft is twisted mediolaterally into a slight S-shape, a situation that mirrors that of their femur in that the proximal and distal joint surfaces of loris tibiae lie in different planes of orientation relative to each other. Galago tibiae are unique

in possessing a prominent bony extension proximally for the articulation of the fibula. Compared to more quadrupedally oriented moving primates, frequent leaping primates exhibit more proximal insertions of the semitendinosus, the gracilis, and the sartorius on the proximal tibia.

FIBULA

The primate fibula is a long and thin bone that articulates with the tibia proximally and distally, leaving a large gap between these two bones for just about their

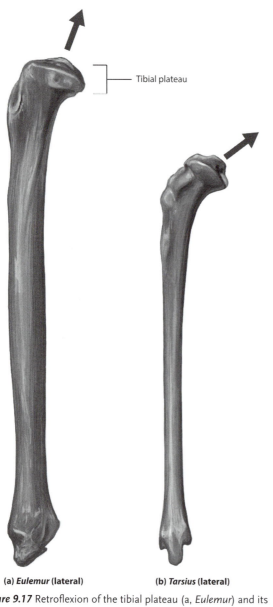

(a) *Eulemur* **(lateral)**   **(b)** *Tarsius* **(lateral)**

Figure 9.17 Retroflexion of the tibial plateau (a, *Eulemur*) and its extreme version (*Tarsius*, b).

Although a fused fibula is rare across living primates, occurring only in *Tarsius*, this fused condition does appear in the Eocene fossil microchoerine called *Necrolemur*. The fibula is also closely apposed in several living primates, including *Microcebus*, and in the callitrichines, as it is in *Apidium phiomense*, a fossil primate from the Oligocene of Egypt. A closely apposed fibula, not to mention a fused one, reduces mobility at the upper ankle joint in these living and fossil primates. Lack of mobility corresponds to a greater use of fore and aft movements at the ankle for leaping in these living primates. We can infer a similar functional explanation for fossil primates with this anatomical condition.

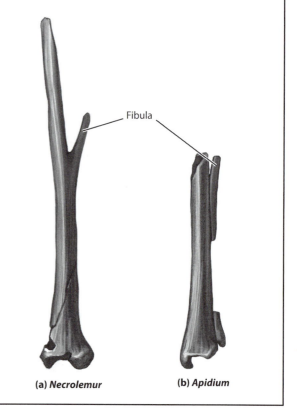

(a) *Necrolemur*   **(b)** *Apidium*

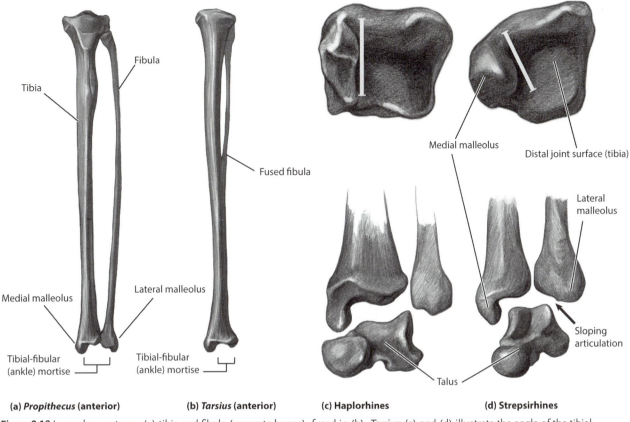

Tibia

Fibula

Fused fibula

Medial malleolus

Lateral malleolus

Tibial-fibular
(ankle) mortise

Tibial-fibular
(ankle) mortise

Medial malleolus

Distal joint surface (tibia)

Lateral
malleolus

Sloping
articulation

Talus

(a) *Propithecus* (anterior)   **(b) *Tarsius* (anterior)**   **(c) Haplorhines**   **(d) Strepsirhines**

Figure 9.18 Lower leg anatomy. (a) tibia and fibula (separate bones), fused in (b), *Tarsius*; (c) and (d) illustrate the angle of the tibial malleolus (*above*) and the anatomical arrangement of the tibia-fibula mortise (upper ankle joint, *below*). Note the oblique angulation of the tibial malleolus and the sloping articulation between the distal fibula and the lateral joint surface of the talus among strepsirhine primates relative to haplorhines.

entire lengths. The fibula is always a separate bone in primates, with the exception of tarsiers, in which this bone is separate for only the proximal third of its length, being completely fused distally (fig. 9.18). Primate fibulae are closely apposed to the distal tibiae in taxa such as *Microcebus, Galago, Galagoides, Callithrix*, and *Cebuella*. The fibulae are much wider in apes compared to the thinner fibulae of cercopithecids. The distal fibula articulates with the lateral facet of the talus in primates. This joint is quite obliquely oriented among the strepsirhine primates, in comparison to the straight joint condition among haplorhines.

THE TIBIAL-FIBULAR MORTISE

The distal tibia and the distal fibula form a mortise to connect the foot to the lower leg (fig. 9.18). This two-prong notch, one prong from the tibia, the tibial malleolus, and one prong from the fibula, the fibular malleolus, grips the medial and lateral sides of the ta-

lus while articulating the talar trochlea with the distal facet of the tibia, thereby forming the upper ankle joint. This tibial-fibular mortise takes three anatomical forms across primates. Among strepsirhines, the medial malleolus of the tibia is long and large, extending distally well past the oblique fibular malleolus. In tarsiers, the medial malleolus extends below the fibular malleolus, but both are parallel in their orientation to each other. Among anthropoids, the medial and lateral malleoli extend to a relatively equal distance across this mortise and both malleoli are parallel to each other. The medial malleolus is shorter and wider in anthropoids relative to that of other primates.

Besides the length of these malleoli in primates, a plantar view shows two distinct morphologies in the two suborders. In strepsirhine primates, the medial malleolus is twisted and highly angled relative to the distal tibial joint surface, whereas in haplorhines the tibial malleolus is perpendicular in its orientation to the distal tibial joint surface (fig. 9.18). This twisting

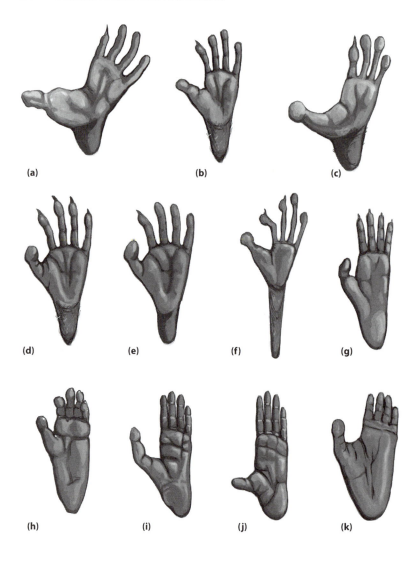

Figure 9.19 Primate feet: (a) *Loris*; (b) *Cheirogaleus*; (c) *Indri*; (d) *Daubentonia*; (e) *Eulemur*; (f) *Tarsius*; (g) *Saimiri*; (h) *Papio*; (i) *Hylobates*; (j) *Pongo*; and (k) *Gorilla*.

among strepsirhines has clear functional implications for enhanced lateral foot mobility relative to haplorhine primates.

The upper ankle joint morphology of strepsirhine and haplorhine primates is distinct (fig. 9.18). In strepsirhines, the distal fibula flares obliquely, making an asymmetrical tibial-fibular mortise, whereas in haplorhines this joint surface is symmetrical with the medial malleolus of the tibia. This mortise anatomy is mirrored in the strepsirhine foot in that the lateral fibular facet of the talus is angled obliquely outward as well. This oblique mortise angle contributes to the lateral rotation of the foot and is a morphology that is especially useful on vertical supports. Among haplorhines, the lateral talofibular facet is steep-sided, like a wall, setting a limit to lateral rotation.

FEET

With the exception of humans, all primates have grasping feet (fig. 9.19). In fact, an opposable big toe, the hallux, is a key innovation for the evolution of primate arboreality. Some primates, like galagos and tarsiers, have long feet with long tarsal bones (i.e., the calcaneus and the navicular; fig. 9.20). This increases the length of the lower limb and improves the mechanics of leaping in these primates.

The primate foot, like the primate hand, is composed of many bones: tarsals, metatarsals, and phalanges (fig. 9.20). The tarsals include the talus, the calcaneus, the navicular, the cuboid, and three cuneiforms (the ento-, meso-, and ectocuneiform). In the wrist, the carpal bones, the equivalent elements to the tarsals of the foot, are aligned in two rows. In the foot, this is

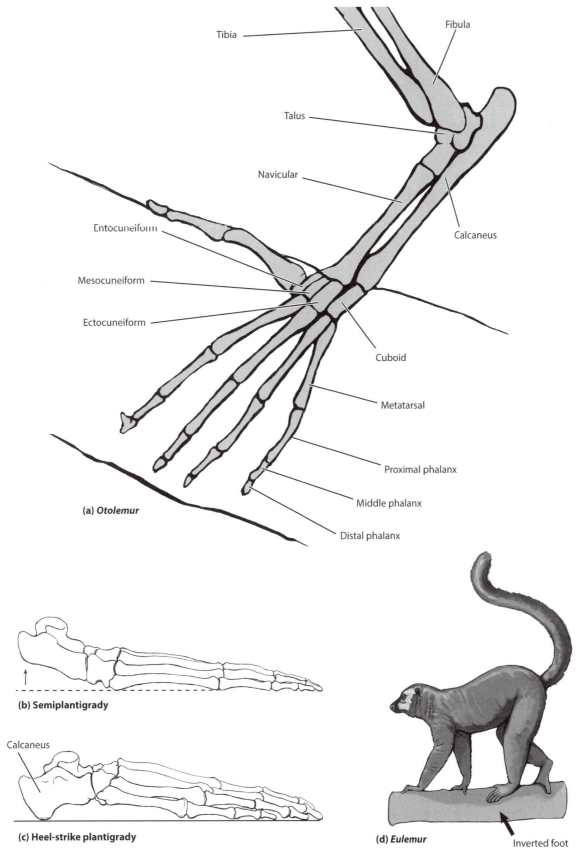

Tibia

Fibula

Talus

Navicular

Calcaneus

Entocuneiform

Mesocuneiform

Ectocuneiform

Cuboid

Metatarsal

Proximal phalanx

Middle phalanx

Distal phalanx

(a) Otolemur

(b) Semiplantigrady

Calcaneus

(c) Heel-strike plantigrady

(d) Eulemur

Inverted foot

Figure 9.20 Primate foot anatomy

not the case. One tarsal bone, the talus, sits on top of the calcaneus and fits into the tibial-fibular mortise. This talar elevation forces the navicular bone, which articulates with the talar head, to be angled relative to the cuboid, the lateral tarsal bone that articulates with the calcaneus at the calcaneocuboid joint. The cuboid and the three cuneiforms articulate with each other in a row similar to that of the carpals. The cuneiforms all articulate with the five metatarsals. The entocuneiform articulates with the first metatarsal, and this joint surface is essential for hallucal grasping. The second and third metatarsals articulate with the mesocuneiform and the ectocuneiform, respectively, leaving the fourth and fifth metatarsals to articulate with the cuboid distally. The proximal joint surfaces of the proximal phalanges all articulate with the metatarsal heads. Like the hand, there are two phalanges, a proximal phalanx and a distal phalanx, within the first digit, whereas three phalanges, proximal, middle, and distal, occur in the four lateral digits.

All primates, with the exception of great apes and humans, use a heel-elevated foot position called semiplantigrady when moving (fig. 9.20). In this foot position the mid-foot contacts the support first. In African apes and humans, the lateral side of the heel, the calcaneus, makes first contact with the support followed by the mid-foot; this type of foot posture is called heel-strike plantigrady (fig. 9.20). This heel up or heel down distinction in primates is important since each contact pattern largely dictates how primate feet work mechanically. Once a foot bone has contacted a support, its orientation affects the compressive forces it must support, allowing some joint surfaces freedom of movement while others are held firm. Foot mobility on curved branches, for example, requires great joint mobility. Among semiplantigrade foot falls, the cuboid and navicular bones support the weight of the leg, allowing the transverse tarsal joint to freely rotate. In contrast, among the heel-strike plantigrade primates, it is the base of the calcaneus that supports weight with an upwardly elevated and largely immobile calcaneocuboid joint.

The key foot joint for primate foot mobility is the transverse tarsal joint. This joint is in fact two joints: the calcaneocuboid joint and the talonavicular joint. These two joints, along with the subtalar joint, provide the rotational ability necessary to place the foot in an inverted position (see chapter 4). Primates turn their feet inward (inversion) to hold on to a horizontal or vertical circular support while the body is aligned sagittally to the support (see fig. 9.20). Grasping works best on small supports, like the terminal branches of a tree, relative to the size of the foot. In contrast, terrestrial primates tend to reduce mobility in their foot joints in favor of stability.

As noted above (see fig. 9.18), the strepsirhine upper ankle joint allows greater lateral rotation of an inverted foot, which is especially helpful on vertical supports. In contrast, the haplorhine upper ankle joint is adapted for foot movements in a fore and aft alignment. Haplorhine primates, however, possess all of the key foot mobility features at the transverse tarsal and subtalar joints for foot inversion and eversion, making them capable climbers when using vertical supports. Haplorhine feet do not possess any special morphological changes that enhance lateral rotation of their feet for vertical support use (e.g., vertical clinging or vertical climbing capabilities).

Strepsirhine primates have shifted the groove for the flexor hallucis longus tendon laterally on the posterior talus, relative to the midline position of haplorhines. This morphological change in strepsirhines also suggests increased use of vertical supports, as do many other modifications around the tibial-fibular mortise.

Oddities

The oddest foot fall pattern is the one noted for orangutans in which the lateral side of their foot contacts the ground first. Orangutans often walk on the lateral sides of their feet with their highly curved toes facing inward or medially. In this way, orangutan foot falls resemble the heel-strike plantigrade pattern of African apes and humans more than the foot fall pattern of primates that utilize semiplantigrady.

FOOT FORM

Foot form is highly varied across primates. Galagos and tarsiers are known for their exceptionally long feet with greatly elongated calcanei and navicular bones. Galagos have even added a mid-navicular contact facet, a synovial joint, with the calcaneus. Tarsiers

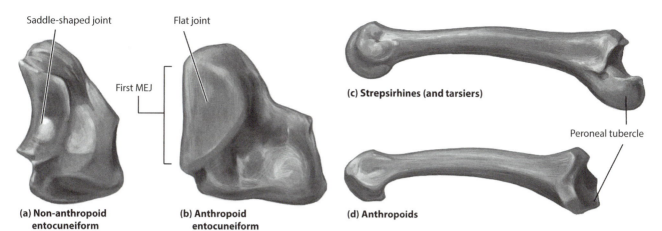

(a) Non-anthropoid entocuneiform

(b) Anthropoid entocuneiform

(c) Strepsirhines (and tarsiers)

(d) Anthropoids

Figure 9.21 Key big toe grasping bones: entocuneiform (a, b) and first metatarsal (c, d). Note that in (a) and (c), the first metatarsal-entocuneiform joint (MEJ) is saddle-shaped (a) and the peroneal tubercle is long (c), in contrast to the flattened joint surface (b) and shortened tubercle (d) of anthropoids.

simply have long, tubular-shaped naviculars and calcanei. In contrast to the long feet of tarsiers and galagos, lemurs and indriids have arched and reoriented their third through fifth digits, making a folded foot rather than the generally flat-lying feet of primates. Lorises have completely reoriented their first digits, having aligned them across from the third digit, and reduced their second digits; both features mimic the odd features of their hands (see fig. 9.19). In fact, most of the bones of the lorisine foot have unusual shapes due to the need to increase mobility throughout the body for acrobatic body and limb motions. In a different adaptive direction, Old World monkeys possess highly asymmetrical talar trochleas with twisted talar heads; their calcanei possess two separate facets distally and a short posterior calcaneal facet. All of these features limit foot mobility and suggest a historical past of terrestriality for all cercopithecids, even the secondarily arboreal ones alive today.

GRASPING TOES

Primate grasping requires flexing of the four lateral digits at the metatarsal-phalangeal and interphalangeal joints while the big toe opposes the lateral digits. The big toe swings across and flexes to form the opposable grasp. This joint motion of opposability is accomplished at the first metatarsal-entocuneiform joint, a saddle-shaped joint that is oriented vertically across non-anthropoids but that is much flatter in anthropoids (fig. 9.21). Terrestrial monkeys, like ba-

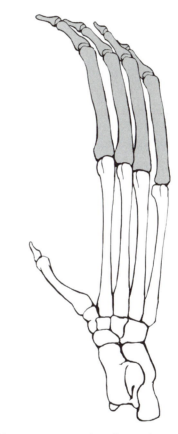

Figure 9.22 Long suspensory adapted toes of an orangutan foot.

boons, tend to de-emphasize their grasping structures and have shorter digits and smaller big toes compared to those of more arboreal primates. In contrast to the terrestrial species, suspensory-oriented primates possess long and highly curved pedal digits, with the best example being the feet of orangutans (fig. 9.22).

Although all five of the digits grasp in primate feet, the grasping big toe, or first digit, is the key. First digits are robust in non-anthropoid primates. Strepsirhines and tarsiers possess a large and robust first metatarsal with a long peroneal tubercle (see fig. 9.20), whereas in anthropoids the first metatarsal is smaller with small peroneal tubercles. From this bone alone, one might surmise that anthropoid grasping is less developed relative to that of the other primates. The oddest big toe arrangement is found in the lorises; their grasping big toe has shifted even farther away from the four lateral digits, orienting itself opposite the third digit, making a more clamping-oriented grasping foot. Their second digit is also highly reduced after this shift.

NAILS

All primates have nailed digits (fig. 9.23), but some taxa have secondarily evolved keeled nails (e.g., *Euoticus, Tarsius pumilus, Phaner*) or claw-like nails on their lateral digits (e.g., *Daubentonia* and callitrichines). Nailed digits mean broader toes, and broad and flattened distal phalanges, with biomechanically wider frictional surfaces to grasp with. Why should primates convert their ancestral mammalian claws to nails? This is a difficult question to answer at present since many small clawed marsupials seem quite efficient on terminal branches, the adaptive domain of primates. Perhaps nails are related to hand use in the act of obtaining and bringing food items to the mouth; or perhaps wider finger or toe tips allow for better adhesion to small, curved branches. Primates may simply have a different style of arboreality relative to other arboreal mammals.

Although *Daubentonia* and *Tarsius pumilus* have pointed or claw-like nails for different adaptive reasons, most of the other "clawed" primates eat gums. Gumnivory is associated with *Phaner, Euoticus*, and the ancestral condition for callitrichids. Claws, or claw-like nails, are easy to explain since these pointed anatomical structures function to puncture and to cling to larger-diameter arboreal supports. As supports increase in diameter, claws allow primates and other mammals like squirrels to cling to the bark surface. In contrast, the grasping span of a primate hand or foot is eventually exceeded as tree diameters increase.

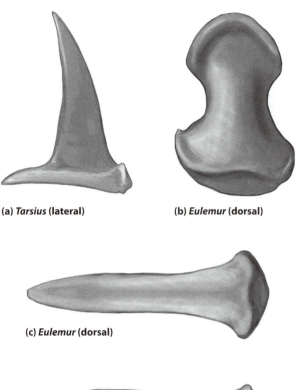

(a) *Tarsius* (lateral)   **(b) *Eulemur* (dorsal)**

(c) *Eulemur* (dorsal)

(d) *Eulemur* (lateral)

Figure 9.23 The distal phalanx for primate digits with grooming claws or nails. A grooming claw from digit two in *Tarsius* (a) relative to a flat nail in *Eulemur* (b). In all living strepsirhines digit two has a grooming claw—dorsal (c) and lateral (d) view, *Eulemur*.

Grasping force is insufficient to hold on to what is essentially a flat wall for nailed primates, and they, of course, fall. Pointed digits overall improve the ability to cling to larger-diameter vertical supports, quite an advantage given the small size of primate hands and feet.

Strepsirhine primates possess a single grooming claw on their second digit, while tarsiers have two on their second and third digits, respectively (see fig. 9.19). No grooming claws are found on anthropoid primates with perhaps the exception of the second digit in *Aotus*. Tarsiers have nails but they also have large, round toe pads, an unusual pad shape among primates.

The distal phalanges of primates are generally flat with wide apical tufts distally. They often show fossae, or depressions, on the first digits proximoplantarly for

Nails are one of the hallmark adaptations for grasping hands and feet across all primates, living and extinct. This anatomical feature is easy to view among living primates. Nails can be inferred from the anatomy of flattened distal phalanges for a host of fossil primates, including some of the earliest known. Illustrated are two distal phalanges from the Eocene adapiform *Notharctus tenebrosus* and the omomyid *Hemiacodon gracilis*, relative to the living *Eulemur fulvus* from Madagascar. Note that all are flattened and arrowheaded in shape, implying nails instead of claws for these two fossil primates. This resemblance in phalangeal morphology suggests a similarity in grasping function for these ancient fossil primates to that in living lemurs like *Eulemur*.

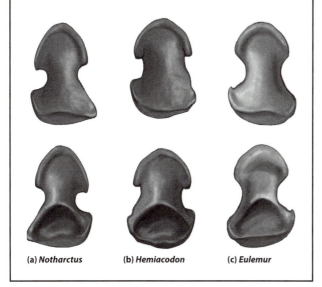

(a) Notharctus   **(b) Hemiacodon**   **(c) Eulemur**

ges, represented by a wider and flatter joint surface. In contrast, metatarsal heads are quite narrow relative to metacarpals, and thus the proximal phalangeal joint is more curved mediolaterally and narrower to reflect metatarsal head shape.

Oddities

Orangutans have highly inverted and curled feet when they walk quadrupedally on the ground. Their long, curved toes are better utilized in suspensory or hook-like grasps. Their curled fingers (fist walking) and gait sequence along the lateral sides of their feet make them an ungainly terrestrial quadruped. Orangutans obviously do the best they can when moving on the ground given their suspensory adapted hands and feet. Orangutan tarsal bones are small relative to their body size, indicating a reduced need to respond to ground reaction forces by their feet. These are feet adapted for suspensory and tensile environment situations, a complete contrast to human foot anatomy and bipedalism.

Lorises have strange-looking feet with their big toes stuck out at an unusual angle, as do indriids, which possess a wide span for their grasping big toes although not to the extreme of a loris. Several primates have clawed rather than nailed lateral digits (e.g., *Daubentonia*).

Selected References

Aiello, L., and C. Dean. 1990. An Introduction to Human Evolutionary Anatomy. Academic Press, New York.

Anemone, R.L. 1990. The VCL hypothesis revisited: patterns of femoral morphology among quadrupedal and saltatorial prosimian primates. American Journal of Physical Anthropology 83:373–393.

———. 1993. The functional anatomy of the hip and thigh in primates; pp. 150–174 *in* D.L. Gebo (ed.), Postcranial Adaptation in Nonhuman Primates. Northern Illinois University Press, DeKalb.

Anemone, R.L., and H.H. Covert. 2000. New skeletal remains of *Omomys* (Primates, Omomyidae): functional morphology of the hindlimb and locomotor behavior of a Middle Eocene primate. Journal of Human Evolution 38:607–633.

Anemone, R.L., and B.A. Nachman. 2003. Morphometrics, functional anatomy, and the biomechanics of locomotion among tarsiers; pp. 97–120 *in* P.C. Wright, E.L. Simons, and S. Gursky (eds.), Tarsiers. Rutgers University Press, New Brunswick, N.J.

the insertion of the flexor tendon. The proximal and middle phalanges of primates are generally elongated and slightly curved anteroposteriorly. These phalanges often show raised medial and lateral edges for flexor sheaths, the connective tissue that keeps digital flexor tendons close to the plantar bony phalangeal surface. The proximal joint surface for the proximal phalanx has a single joint surface for contact with a metatarsal head, while the middle phalanges possess a double joint surface for articulation with the distal joint surface of a proximal phalanx. It is difficult, perhaps impossible, to distinguish manual from pedal distal or middle phalanges. Proximal phalanges can be sorted, however. There are several subtle indicators, but the best may be the proximal joint surface that contacts a wide metacarpal head for the manual proximal phalan-

Ankel-Simons, F. 2000. Primate Anatomy—An Introduction. Academic Press, New York.

Beard, K.C., M. Dagosto, D.L. Gebo, and M. Godinot. 1988. Interrelationships among primate higher taxa. Nature 331:712–714.

Cant, J.G.H., D. Youlatos, and M.D. Rose. 2001. Suspensory locomotion of *Lagothrix lagothricha* and *Ateles belzebuth* in Yasuni National Park, Ecuador. Journal of Human Evolution 44:685–699.

Cartmill, M. 1972. Arboreal adaptations and the origin of the Order Primates; pp. 97–122 in R.H. Tuttle (ed.), The Functional and Evolutionary Biology of Primates. Aldine Press, Chicago.

———. 1974. Pads and claws in arboreal locomotion; pp. 45–83 in J.A. Jenkins (ed.), Primate Locomotion. Academic Press, New York.

———. 1979. The volar skin of primates: its frictional characteristics and their functional significance. American Journal of Physical Anthropology 50:497–510.

———. 1985. Climbing; pp. 73–88 in M. Hildebrand, D.M. Bramble, K.F. Liem, and D.B. Wake (eds.), Functional Vertebrate Morphology. Belknap Press of Harvard University Press, Cambridge, MA.

Cartmill, M., P. Lemelin, and D.O. Schmitt. 2002. Support polygons and symmetrical gaits in mammals. Zoological Journal of the Linnean Society 136:401–420.

Crompton, R.H., W.I. Sellers, and M.M. Gunther. 1993. Energetic efficiency and ecology as selective factors in the salutatory adaptation of prosimian primates. Proceedings of the Royal Society, Series B. 254:41–45.

Dagosto, M. 1983. Postcranium of *Adapis parisiensis* and *Leptadapis magnus* (Adapiformes, Primates). Folia Primatologica 41:49–101.

———. 1985. The distal tibia of primates with special reference to the Omomyidae. International Journal of Primatology 6:45–75.

———. 1990. Models for the origin of the anthropoid postcranium. Journal of Human Evolution 19:121–139.

Dagosto, M., and D.L. Gebo. 1994. Postcranial anatomy and the origin of the Anthropoidea; pp. 567–592 in J.G. Fleagle and R.F. Kay (eds.), Anthropoid Origins. Plenum Press, New York.

Dagosto, M., and P. Schmid. 1996. Proximal femoral anatomy of omomyiform primates. Journal of Human Evolution 30:29–56.

Dagosto, M., D.L. Gebo, and K.C. Beard. 1999. Revision of the Wind River faunas, Early Eocene of central Wyoming. Part 14. Postcranium of *Shoshonius cooperi* (Mammalia: Primates). Annals of Carnegie Museum 68(3):175–211.

Dagosto, M., D.L. Gebo, X. Ni, T. Qi, and K.C. Beard. 2008. Primate tibiae from the Middle Eocene Shanghuang Fissure-fillings of Eastern China; pp. 315–324 *in* E.J. Sargis and M. Dagosto (eds.), Mammalian Evolutionary Morphology: A Tribute to Frederick S. Szalay. Springer, Dordrecht, Netherlands.

Demes, B., W.L. Jungers, J.G. Fleagle, R.E. Wunderlich, B.G. Richmond, and P. Lemelin. 1996. Body size and leaping kinematics in Malagasy vertical clingers and leapers. Journal of Human Evolution 31:367–388.

Demes, B., J.G. Fleagle, and P. Lemelin. 1998. Myological correlates of prosimian leaping. Journal of Human Evolution 34:385–399.

Demes, B., J.G. Fleagle, and W.L. Jungers. 1999. Takeoff and landing forces of leaping strepsirhine primates. Journal of Human Evolution 37:279–292.

Fleagle, J.G., and F. Anapol. 1992. The indriid ischium and the hominid hip. Journal of Human Evolution 22:285–305.

Fleagle, J.G., and E.L. Simons. 1979. Anatomy of the bony pelvis in parapithecid primates. Folia Primatologica 31:176–186.

Fleagle, J.G., and E.L. Simons. 1983. The tibio-fibular articulation in *Apidium phiomense*, an Oligocene anthropoid. Nature 301:238–239.

Fleagle, J.G., and E.L. Simons. 1995. Limb skeleton and locomotor adaptations of *Apidium phiomense*, an Oligocene anthropoid from Egypt. American Journal of Physical Anthropology 97:235–289.

Gebo, D.L. (ed.). 1993. Postcranial Adaptation in Nonhuman Primates. Northern Illinois University Press, DeKalb.

———. 1989. Postcranial adaptation and evolution in Lorisidae. Primates 30(3):347–367.

Gebo, D.L., M. Dagosto, K.C. Beard, and X. Ni. 2008. New hindlimb elements from the Middle Eocene of China. Journal of Human Evolution 55:999–1014.

Grand, T.I. 1967. The functional anatomy of the ankle and foot of the slow loris (*Nycticebus coucang*). American Journal of Physical Anthropology 28:168–182.

Gregory, W.K. 1920. On the structure and relations of *Notharctus*, an American Eocene primate. Memoirs of the American Museum of Natural History 3:49–243.

Hall-Craggs, E.C.B. 1965. An analysis of the jump of the lesser galago. Journal of Zoology (London) 147:20–29.

Hamrick, M.W. 1998. Functional and adaptive significance of primate pads and claws: evidence for New World anthropoids. American Journal of Physical Anthropology 106:113–127.

———. 2001. Primate origins: evolutionary change in digital ray patterning and segmentation. Journal of Human Evolution 40:339–351.

Hartman, C.L., and W.L. Straus (eds.). 1971. The Anatomy of the Rhesus Monkey. Hafner Publishing Company, New York.

Jouffroy, F.K., and J. Lessertisseur. 1979. Relationships between limb morphology and locomotor adaptations among prosimians: an osteometric study; pp. 143–182 *in* M.E. Morbeck, H. Preuschoft, and N. Gomberg (eds.), Environment, Behavior, and Morphology: Dynamic Interactions in Primates. Gustav Fischer, New York.

Jungers, W.L. 1985. Body size and scaling of limb proportions in primates; pp. 345–382 in W.L. Jungers (ed.), Size and Scaling. Plenum Press, New York.

Kapanji, I.A. 1987. The Physiology of the Joints, Vol. 2: Lower Limb. 5th edition. Churchill Livingstone, Edinburgh.

Larson, S.G. 1998. Unique aspects of quadrupedal locomotion in nonhuman primates; pp. 157–173 in E. Strasser, J.G. Fleagle, A.L. Rosenberger, and H.M. McHenry (eds.), Primate Locomotion: Recent Advances. Plenum Press, New York.

Larson, S.G., and J.T. Stern. 2009. Hip extensor EMG and forelimb/hindlimb weight support asymmetry in primate quadrupeds. American Journal of Physical Anthropology 138:343–355.

Larson, S.G., D. Schmitt, P. Lemelin, and M. Hamrick. 2001. Limb excursion during quadrupedal walking: how do primates compare to other mammals? Journal of Zoology (London) 255:353–365.

Lewis, O.J. 1989. Functional Morphology of the Evolving Hand and Foot. Oxford Science Publications, Oxford, UK.

McArdle, J.E. 1981. The Functional Morphology of the Hip and Thigh of the Lorisiformes. Contributions to Primatology, Vol. 17. Karger, Basel.

Morton, D.J. 1924. The evolution of the human foot, part II. American Journal of Physical Anthropology 1:1–52.

Ruff, C.B., and J.A. Runestad. 1992. Primate limb bone structural adaptations. Annual Review of Anthropology 21:407–433.

Schultz, A.H. 1969. The Life of Primates. Weidenfeld and Nicolson, London.

Szalay, F.S., and M. Dagosto. 1988. Evolution of hallucial grasping in the primates. Journal of Human Evolution 17:1–33.

Vilensky, J.A. 1989. Primate quadrupedalism: how and why does it differ from that of typical quadrupeds? Brain, Behavior and Evolution 34:357–364.

Ward, C.V. 1993. Torso morphology and locomotion in Proconsul nyanzae. American Journal of Physical Anthropology 92:291–328.

White, J.L., and D.L. Gebo. 2004. A unique proximal tibia morphology in strepsirhine primates. American Journal of Primatology 64:293–308.

10 | Great Ape and Human Anatomy

Edward Tyson in 1699 was the first individual, at least in print, to recognize that ape anatomy was similar to human anatomy. Tyson's dissections compared a monkey, an ape, and a human, and he recognized that apes and humans were anatomically more similar to each other than either was to a monkey. Linnaeus would later come to the same conclusion and group apes and humans together in his *Systema Naturae* in 1758. Although neither believed in evolution, nor were they alive to read Charles Darwin's 1859 *The Origin of Species*, the recognition of anatomical similarities shared by apes and humans would alter our perspective on humanity forever.

The phyletic position of the ape lineages relative to humans is clear today. Gibbons, an Asian ape lineage, represent the most primitive hominoid alive currently relative to the great ape and human lineages (fig. 10.1). Orangutans, also from Asia, are the next phyletic outgroup. Both gibbons and orangutans are more primitive apes relative to us in regard to our evolutionary closeness with African apes. Of the three living African apes, gorillas represent the first outgroup and are the least closely related to humanity. Chimpanzees and bonobos are our closest living kin; together they represent our sister taxa, including all of the human fossil record. The ape cladogram illustrated in figure 10.1 shows humanity to be more closely related to chimpanzees than chimpanzees are to gorillas—an odd idea when we consider the anatomical differences between chimpanzees and humans relative to the

anatomical similarities chimpanzees share with gorillas. Nevertheless, this is a true fact. In the end, human origins, like that of hominoids and catarrhines, are originally African, and therefore we are all Africans under the skin.

HEADS

The heads of great apes have smaller cranial capacities and larger facial regions relative to human skulls (fig. 10.2). Great ape brains are large relative to their body size, but only about one-third in absolute size relative to modern human brain sizes. Human head shape is quite distinctive compared to that of great apes in that it is rather globular or round in lateral view relative to the long skulls in apes, which possess better-developed rostrums and faces. In contrast, humans have small faces that have shifted back underneath the braincase. A short and small human face has little room left for the prognathic nasal region or the large anterior teeth of apes.

The tall forehead observed in human skulls and the severe bending at the cranial floor are diagnostic characters of the human skull relative to ape skulls. Likewise, the cranial base is short and wide in humans compared to the long and narrow condition in apes. The subnasal morphology within human skulls is similar to that of African apes, however, where it forms a step-down pattern between the premaxilla and the maxilla (fig. 10.3). This subnasal region is

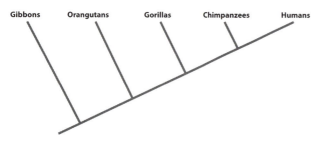

Figure 10.1 Ape and human cladogram.

particularly unique in orangutans, where no step exists and the incisive canal is narrow (fig. 10.3).

African ape and human faces both have squared orbit shapes with a broad interorbital region. In contrast, orangutan skulls are unique in that they possess oblong or tall orbits, a narrow space between the orbits (the interorbital region), and a dished face from the side, sometimes described as strong dorsal flexion of the midface, or airorhynchy, relative to the facial profile of African apes and humans.

Great ape and human skulls are also known for their sinuses. The maxillary sinus is large in all of the great apes and humans and all possess a sphenoid sinus as well. Orangutan skulls lack an ethmoid sinus and a frontal sinus in contrast to the skulls of African apes and humans.

The occipital bone at the back of the human skull shows the entry position for the foramen magnum. The foramen has shifted anteriorly, allowing the human head to balance on an erect vertebral column, being similar to a ball on top of a pole. In contrast, ape heads are more similar to anthropoid skulls, where the foramen magnum is positioned posteriorly.

The large nuchal region at the back of great ape skulls is another region of stark contrasts to human occipitals. This region is where the large neck muscles attach to hold up the enlarged heads of apes, especially the massive heads of male gorillas. Gorillas also possess tall spinous processes along their cervical vertebrae for the attachment sites of these large neck muscles in comparison to the smaller human neck musculature and small cervical vertebrae and spinous processes (fig. 10.4).

In contrast to human skulls, great ape skulls are known for their robust masticatory system with large attachment areas for the three chewing muscles (the temporalis, masseter, and pterygoid muscles). Great ape skulls commonly have browridges, sagittal crests

(especially in males), and large cheek regions (figs. 10.2 and 10.5). In contrast, the bizygomatic width (or mid-facial width) is reduced in human skulls and none possess large muscle crests. Gorilla heads have an especially large supraorbital torus above the eyes with a sulcus behind this torus. Muscle cresting is again especially prominent along the skulls of male gorillas. Besides the large muscle attachment surfaces, the zygomatic bone is located more posteriorly, along the side and behind the medial orbital rim in humans, relative to this bone's more forward-positioned cheeks and nasal regions among the great apes.

The temporomandibular joint also differs between great apes and humans. This joint surface is shallow in ape skulls but shows a deep depression in human temporal bones. The entoglenoid process is large, the postglenoid process is long, and the mandibular condyle fits closely within the glenoid fossa in great apes, thereby restricting side-to-side jaw movements (fig. 10.5). Human mandibular joints have more latitude for medial and lateral movements.

The mandibles of apes are large relative to those of humans with especially bony buttresses internally, the superior and inferior mandibular tori, or, as this region is often called, a simian shelf (fig. 10.6). Human mandibles have a chin, a bony buttress on the outside of the symphysis, while this outward anatomical region slopes away in ape mandibles. The mandibular ascending ramus is much wider and longer in apes relative to its smaller version in humans. Gorillas have an especially wide and tall ascending ramus.

TEETH

The most obvious dental feature when examining great ape teeth is their large canines, especially the tall and sharp upper canines, in contrast to the attenuated and blunted canines of humans (fig. 10.6). Humans have greatly reduced their anterior dentition, the incisors and canines, relative to that of the great apes, but we do have large molars relative to our body size (fig. 10.6). We process foods more often with our cheek teeth, whereas great apes, especially the orangutans, frequently use front tooth nipping, stripping, and cutting to process foods. Tooth use and food processing are also evidenced in the thickness of molar enamel; humans have the thickest enamel,

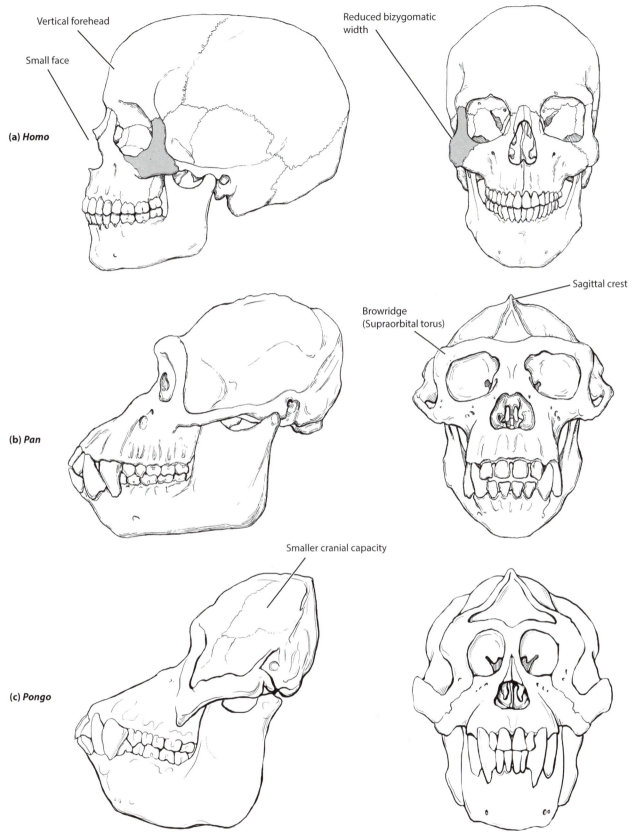

Figure 10.2 Great ape and human skull comparisons.

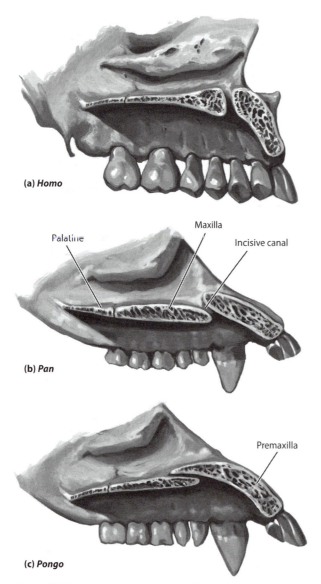

(a) *Homo*

Palatine
Maxilla
Incisive canal

(b) *Pan*

Premaxilla

(c) *Pongo*

Figure 10.3 Subnasal palate morphology in great apes and humans. Note the premaxilla and maxilla bony overlap with a small horizontally oriented incisive canal in orangutans relative to *Pan* and *Homo*.

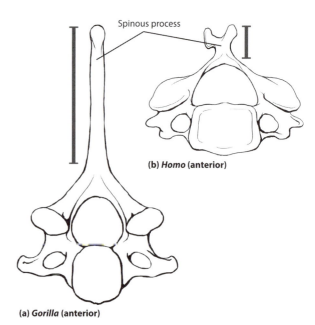

Spinous process

(b) *Homo* (anterior)

(a) *Gorilla* (anterior)

Figure 10.4 Cervical spinous process comparisons between a gorilla (a) and a human (b).

followed by orangutans, and then the thin-enameled African apes. Enamel thickness suggests that ancestral human foods were gritty, abrasive, or hard, requiring extensive molar chewing and thickened enamel helped to prevent dental wear. These abrasive types of food items contrast with the soft fruits or leaves associated with the diet of thin-enameled African apes.

In terms of dental shape, humans possess a parabolic dental arcade relative to the rectangular-shaped palate of great apes (fig. 10.6). This rectangular-shaped palate is particularly noteworthy when we examine the large upper canines of the great apes since the dental arcade must allow these large teeth to fit between the incisors and cheek teeth. Losing our ancestrally large canines altered the human dental arcade since the bones housing these teeth no longer required the extra space to accommodate large canines.

Both great ape and human lower molars have five cusps, having added a hypoconulid, and the Y-5 lower molar cusp pattern is a classic designation for human and ape lower molars (fig. 10.6). Orangutans are noteworthy in their possession of crenulated, or wrinkled, enamel on their molars, while gorillas have taller molar cusps with substantive talonid basins and generally large postcanine teeth relative to the other great apes (fig. 10.6). Gorillas have the largest hypocones among the great apes as well. Human molars are bunodont (having cusps that are separate and rounded), being similar to those of chimpanzees, although overall large in size relative to our anterior dentition, a contrasting pattern in comparison to a chimpanzee dental arcade.

All great apes and humans have large upper central incisors relative to the lateral incisors. In orangutans and gorillas the lateral upper incisors are more pointed, whereas chimpanzees and humans are more similar in I2 morphology, regarding their size and

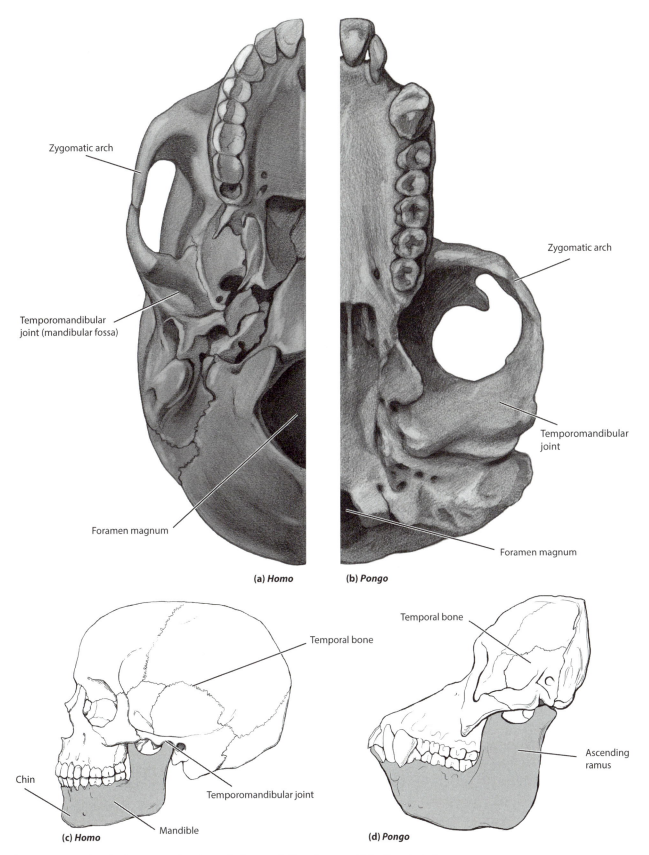

Zygomatic arch

Temporomandibular
joint (mandibular fossa)

Foramen magnum

Zygomatic arch

Temporomandibular
joint

Foramen magnum

(a) *Homo*   **(b) *Pongo***

Temporal bone

Temporal bone

Chin

Temporomandibular joint

Ascending
ramus

Mandible

(c) *Homo*   **(d) *Pongo***

Figure 10.5 Ventral and side comparative views of a human (*left*) and an ape (*right, Pongo*) skull.

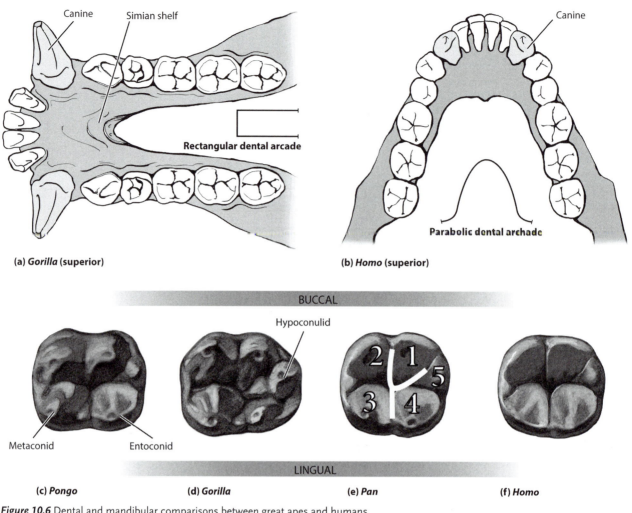

Figure 10.6 Dental and mandibular comparisons between great apes and humans.

general shape, to I1. The lower I2 is larger than I1 in all apes and humans, but this size difference is greatest in gorillas. The upper incisors of orangutans extend even more up and away from the maxilla relative to those of African ape upper incisors (see fig. 10.2).

All of the great apes and humans possess bicuspid upper premolars with larger and taller paracones relative to protocones. These two cusps are more equal in size among orangutan premolars. Lower P3s are sectorial for the great apes relative to humans and lower P4s are also two-cusped (in this case a metaconid and a protoconid) like the two-cusped upper premolars.

FORELIMBS

The arms are long in great apes relative to those of humans, but shoulder, elbow, and wrist morphology, as well as arm function, is functionally similar for all.

A few notable differences include gibbons with especially elongated forelimbs, especially the radius and ulna, relative to great ape forelimbs. Medial torsion is greatest in the humerus of African apes and humans relative to values recorded for the Asian apes. A lower torsion angle for shoulder orientation leaves gibbons with laterally splayed forelimbs in contrast to more forward- or anterior-facing elbows displayed among the great apes and humans.

Perhaps the biggest forelimb differences between apes and humans occur within the hands, where human lateral fingers are short relative to the long fingers of apes. This finger pattern reverses for the human thumb, which is relatively long in comparison to the relatively shorter thumbs among the living apes (fig. 10.7). This proportional distinction between ape and human fingers is indicative of hand function. The long, curved fingers of apes are important for grasping

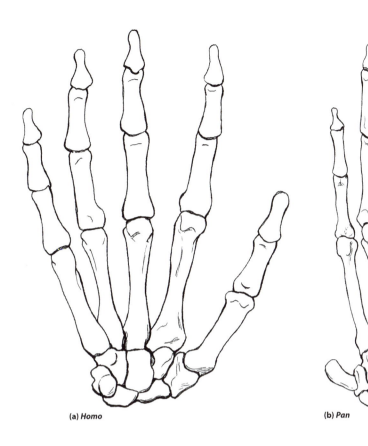

(a) *Homo*   **(b)** *Pan*

Figure 10.7 Chimpanzee versus human hands. Note the elongated lateral digits in chimpanzees relative to the long thumb in humans.

curved branches, whereas human hands are dexterous with elaborate thumb movements. The human thumb is able to touch the tips of each of the four lateral digits, and finger opposition is an important distinction for precision gripping, in contrast to the power grips of ape hands (fig. 10.8). The human thumb is modified, relative to the thumb of the great apes, in having a much bigger flexor tendon that makes a large insertion depression on the distal phalanx of the first digit phalanx, relative to that in apes. Humans have a separate tendon, relative to their lateral digits, for thumb flexion (fig. 10.8). Our finger tips are also broad and dexterous relative to those of apes. In the end, human hands show selection for fine manipulation of objects (i.e., tool use). In contrast, apes, especially the Asian apes, have long lateral fingers, which are best utilized in suspensory grasps or hanging postures and movements.

Another hand feature concerns wrist anatomy. Both African apes and humans display broad wrist bones for weight bearing, the ancestral condition for humanity. Orangutan and gibbon wrists, in contrast, are narrow with greater rotational mobility, an important capability for frequent brachiators.

BODY

As noted in chapter 7, the upper body of great apes and humans is similar in overall construction, being quite different from those of other primates. Orthograde backs, reoriented scapulae, and large joint movements at the shoulder have already been discussed in chapter 8. One distinctive feature is that the human thorax is shaped like that of a gibbon, where the lower end of the thorax tapers inward, in contrast to African apes and orangutans, which show a more widely splayed lower thorax, being funnel-like in overall shape.

In contrast to the many similarities shared by great apes and humans in the construction of the thorax, the human back is a novelty and an adaptation for an erect body posture. Human vertebral bodies are enlarged for weight bearing, the lower back is stiff, and two curves exist along the spine (fig. 10.9). Ape backs maintain greater freedom of movement. Great apes are often arboreal, climbing trees or moving quadrupedally through the canopy, with backs and limbs needing to be placed in several different orientations. Humans stand and walk. This locomotor and postural

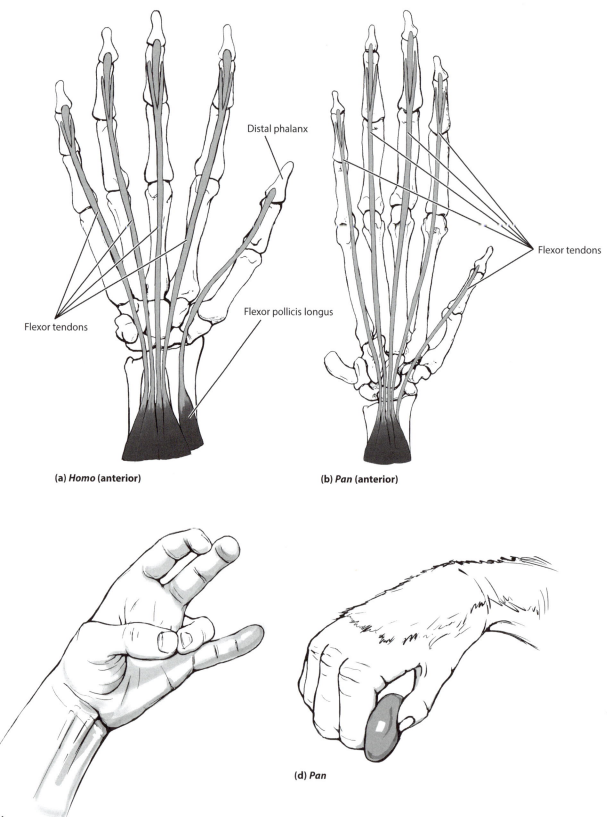

(a) Homo (anterior)

Distal phalanx

Flexor tendons

Flexor pollicis longus

(b) Pan (anterior)

Flexor tendons

(c) Homo

(d) Pan

Figure 10.8 Grasping and muscle insertions of humans and chimpanzees. Note the separate muscle belly and tendon for digit one, the thumb (*top left*), and the digital opposition (*bottom left*) in humans in comparison to chimpanzee fingers.

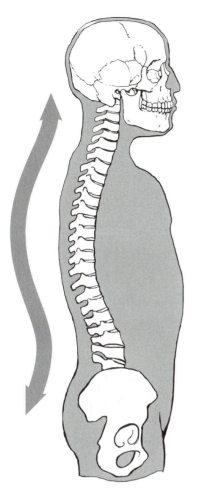

Figure 10.9 Human spine curvature.

distinction is profoundly documented in the bodies of humans.

The human vertebral column is made to bear weight in an erect body plan. This body plan continues downward along the vertebral column to the wide sacrum and the promontory joint surface and the closely tied sacroiliac joints. The promontory joint surface is the last point of bony contact before the upper body weight can be redistributed to each leg (fig. 10.10). The two articular surfaces for the sacrum and the ilium at the sacral-iliac joints are also enlarged in humans, relative to great apes, as is the iliac tuberosity, where all of the ligaments that bind these two bones fit together. The human sacrum is wide and also more angled in comparison to an ape sacrum.

The position of the human pelvis compared to a great ape pelvis shows precisely how the alignment of forces, muscle positions, and modified bone changes have occurred to distinguish these two groups

(fig. 10.10). A shorter lumbar region with a reduced distance between the thorax and the iliac crests in apes implies an immobile lower spine. In humans, the lower spine or lumbar vertebrae are quite mobile relative to those of great apes. Human lower vertebrae are longer, being associated with the shortening and broadening of the human ilia and sacrum. The widening and shortening of the gluteal plane on the ilium and the anteromedial bending of the iliac tubercle brings the smaller gluteal muscles, the gluteus medius and the gluteus minimus, to the side of the human pelvis where they act as abductors. The pelvic tilt mechanism, or lateral balance control, in the human gait is a fundamental innovation for human bipedalism (fig. 10.11). This mechanism allows humans to maintain balance as they swing their pelvis away from the stance leg while walking. The gluteus medius and the gluteus minimus are the muscles responsible for pelvic stabilization during the single support phase of the human gait. Human weight transfer must go from the promontory of the sacrum and the sacroiliac joints to the pelvic acetabulae, to the head and neck of each femur, continuing downward to the foot as humans stand and walk. Weight transfer in human bipedalism contrasts greatly with the multiple-purpose arms and legs of great apes.

The human pelvis has one other key innovation and that is the iliac pillar, which reinforces the ilium relative to the pull of the gluteal muscles. The gluteus maximus stabilizes the trunk on the lower limb, bracing the sacroiliac joint. It helps to keep the human trunk from falling forward as we stand and walk.

Another innovation at the human hip is the possession of massive ligaments (fig. 10.12), the iliofemoral ligament anteriorly and the ischiofemoral ligament posteriorly, in comparison to the hip ligaments of great apes. These two enlarged ligaments help to strengthen the human hip joint given the enlarged femoral head anatomy occurring in humans in contrast to femoral head size in apes. One other distinctive feature occurs internally within the femoral neck and illustrates the force changes that distinguish leg use in humans from leg use in apes. The internal bony architecture in a cross-sectional view is distinctive at the femoral neck region of apes, being asymmetrical, overall thickened with plenty of compact bone, and pointed on one side; in humans the internal architecture of this region

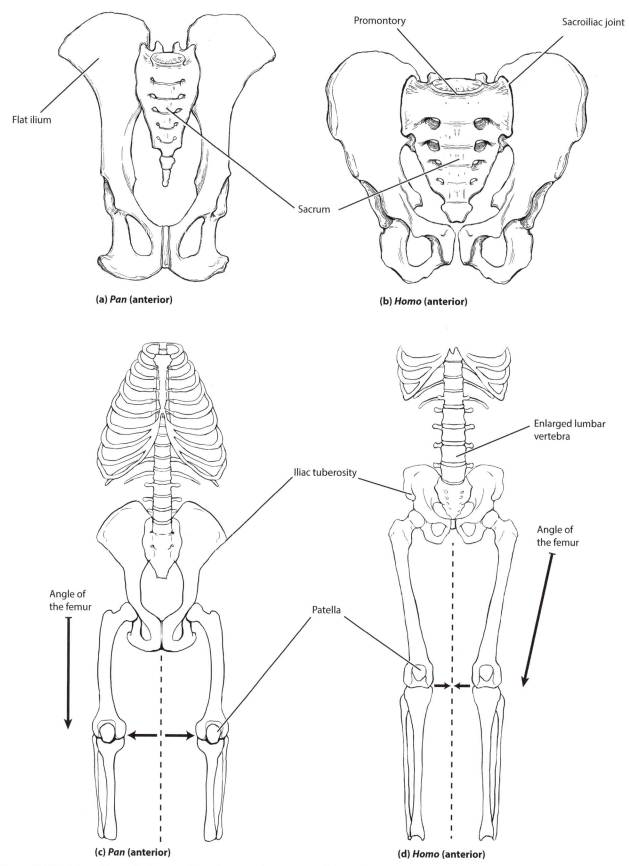

Figure 10.10 Pelvic and lower limb comparisons between chimpanzees (*left*) and humans (*right*).

is ring-like, with a thin band of compact bone surrounding a large ring of trabecular bone with moderate compact bone thickening only on one side.

In the end all of the ligamentous attachments at the hip are critical for human walking and standing postures since the human leg angles inward toward the midline of the body, rather than being oriented straight downward as in the legs of apes. Given our long legs, this knock-knee posture is a crucial distinction between great ape and human knees, as the bicondylar angle alters the angle of patellar movement relative to the hip (see below).

LOWER LIMBS

The lower limbs of humans are dramatically remodeled relative to great ape limbs. Human bipedalism is the hallmark adaptation of humanity and this locomotor adaptation occurs long before big brains or the invention of stone tools. The human pelvis, thigh (femur), knee, lower leg, and foot have changed dramatically from that of ape hindlimbs. First, the human pelvis is bowl-shaped. It is broad rather than long. The human innominate has lost the long and flat

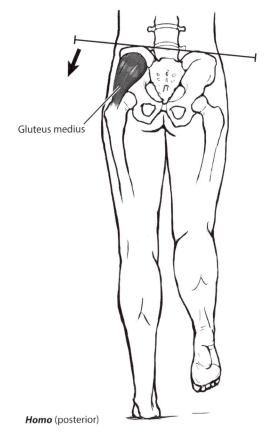

Gluteus medius

Homo (posterior)

Figure 10.11 Pelvic tilt mechanism in human bipedalism.

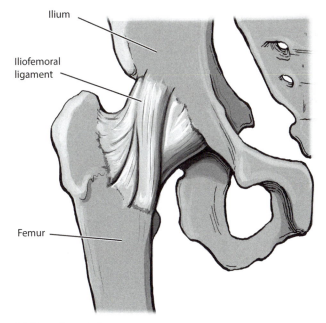

Ilium

Iliofemoral
ligament

Femur

(a) *Homo* (anterior)

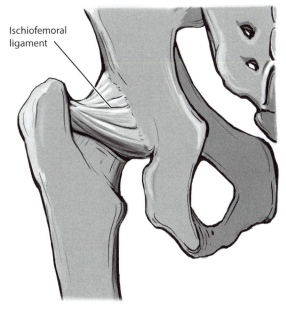

Ischiofemoral
ligament

(b) *Homo* (posterior)

Figure 10.12 Ligaments of the human hip.

ilium of apes (see fig. 10.10). The shortening of the human ilium and ischium moves the muscle insertions closer to the hip joint, sacrificing extra mechanical force (lever length) for hip flexors and extensors for an increase in speed during hip extension. In contrast, apes have a long and flat ilium that lies upward along the lower back. This back orientation helps in leg extension and illustrates how the gluteal muscles are positioned differently in great apes and humans. The gluteus maximus, an important muscle when climbing stairs (hip extension) in humans, is especially large in humans, and it curves along the innominate to attach onto the femur. It occupies a more posterior position of attachment relative to that of great apes. In apes, the gluteus maximus has an extensive insertion along the femur and acts as an abductor.

The gluteus medius, in particular, and the gluteus minimus are important pelvic muscles for maintaining balance while walking (see fig. 10.11). They both act as abductors. When walking, humans first heel-strike, move the leg over the foot, plant the foot at stance phase, and then toe-off. At stance phase, the opposite leg is toeing off and swinging forward for the next stride. This means that at stance phase all of the body weight is situated only on one leg. The human body compensates by tilting the pelvis to one side (i.e., hip sways). The pelvic tilt mechanism in humans is keenly important for maintaining balance while walking, essentially one leg at a time. The gluteus medius and minimus accomplish this balancing act of tilting the pelvis by contracting muscles on the stance leg side to abduct the hip and thorax. The wider the hips (e.g., women), the greater the tilt or sway. Women Olympic sprinters have very narrow hips since this shifts their legs (via their hips) closer to the midline of their body for great mechanical efficiency. The iliac pillar, unique to humans, is a bony buttress along the ilium that adds support for these hip abductors.

The human femur is long and held at an angle from the hip to the knee, as noted above. This carrying angle means that humans are knock-kneed, with the head of the femur above the lateral condyle of the distal femur (see fig. 10.10). Our feet are much closer together than are the knees and feet of great apes. A long femur and overall hindlimb increases a human stride distance relative to that of the short-legged

great apes. The human knee is distinctive in its height and patellar joint surface relative to that of great apes, which possess a mediolaterally wide and flattened knee joint. Humans possess a tall lateral patellar rim relative to the wide and flatter patellar joint surface of ape femora. This bony modification is due to human quadriceps muscle contractile force pulling on the patella during leg extension. In apes, the patella simply moves upward since great apes have straight-oriented legs. In humans, the angle of the femur causes a mechanical problem in that the patella wants to move laterally and out of the patellar groove. To prevent this, humans have evolved a tall bony rim, the lateral patellar rim, and this structure keeps the patella within its groove as it moves superiorly (fig. 10.13).

The distal femoral condyles are flattened at the knee joint in humans, often described as elliptical, relative to the condyles of apes. The lateral femoral condyle is elongated anteroposteriorly, allowing greater cartilage contact during extension in the human knee. The large size of the human tibial epicondyles and

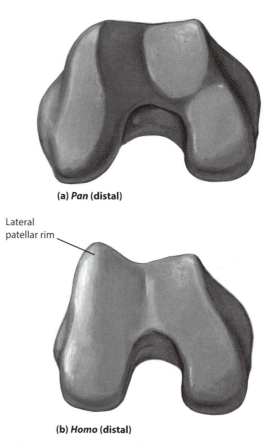

(a) *Pan* (distal)

Lateral patellar rim

(b) *Homo* (distal)

Figure 10.13 Knee joint anatomy of apes (a) and humans (b).

their shape modifications, relative to those of great apes, helps in weight bearing since the human knee can be hyperextended and "locked" in extension for long periods of time when standing, with little muscular effort. The cruciate ligaments, within the femoral notch, help guide the tibia into a position of lateral rotation under the femur. The knee is unlocked using the popliteus muscle along the back of the knee joint.

Distally, the angle of the distal fibular joint surface that articulates with the talus is large in great apes but small in humans. As a consequence, the fibula does not transmit any significant weight to the foot in humans, as it does in great apes. Humans also have a small gap between the tibia and the fibula distally relative to this region in great apes.

As we walk, heel-strike plantigrady positions the human foot and by extension the leg on a terrestrial substrate. The human foot acts as a platform for weight transfer since humans have lost their ancestral grasping big toe (fig. 10.14). Human tarsal bones are large and restricted in their mobility, especially rotationally, while human toes are shortened and small (i.e., not good for grasping). The one exception to this size distinction is the human big toe, which is hyper-enlarged for weight bearing (fig. 10.14). The first digit in humans is aligned next to the second digit and is largely immobile. First digit adduction greatly contrasts with the feet of great apes, which utilize an abducted grasping big toe. The loss of a grasping big toe and the subsequent enlargement of the first digit (the first metatarsal, the first proximal and first distal phalanges) is one of the most profound anatomical distinctions between African apes and humans, being entirely an adaptation for human bipedalism. The loss of opposability is also evidenced by the flattening of the first metatarsal-entocuneiform, joint which stabilizes this joint in order to transmit walking forces instead of providing mobility for an opposable big toe. The size of the human first metatarsal also suggests an increase in force transmission through this digit. The size and robust bony nature of this first digit is evident in the human gait sequence of toe-off when all body weight is borne by the distal phalanx of the first digit as the opposite leg swings forward. The human distal phalanx is also twisted distally with a marked lateral rotation relative to the first distal phalanx of apes.

To add to the platform stability of the human foot, the calcaneocuboid joint locks the calcaneus and cuboid bones together (fig. 10.14). The talar trochlea is also wide for weight bearing at the upper ankle joint and the heel region of the calcaneus has added an extra lateral plantar tubercle for greater width and weight support (fig. 10.14). The human calcaneus is in fact quite robust as a tarsal element. Additional bone bracing for buttressing has been added in several bones of the human foot, and these tarsals are largely locked into place with little movement abilities. Our foot essentially moves at the upper ankle joint (i.e., foot flexion and extension) with little movement capabilities elsewhere among the other joints of the foot.

Human lateral toes are very short and the distal phalanges mere bits of bone relative to the phalanges of apes. Human metatarsals do show hyperextension abilities at the metatarsal-phalangeal joints at toe-off. Human metatarsal heads are expanded along their dorsal articular surfaces with a groove between the head and shaft, allowing hyperextension by the toes (the proximal phalanx) during the toe-off phase of bipedalism (fig. 10.15). This metatarsal region appears more dome-like in humans in comparison to the asymmetrical rounded metatarsal heads of great apes.

The human foot has a longitudinal arch. Here the medial side of the foot, the in-step, is elevated above the ground surface. The spring ligament, a ligament between the calcaneus and navicular that lies below the talar head, and the backward shift of the long plantar ligament, a ligament between the calcaneus and cuboid, are the two key ligament changes to help form this arch (fig. 10.16). The plantar aponeurosis of the foot and several long muscle tendons also help to strengthen the arch. The human medial arch is important in that it absorbs and returns energy to the foot, especially during running.

In contrast to the relatively immobile feet of humans, the feet of great apes are muscular with a grasping big toe (see fig. 10.14). Ape feet are adapted for tree use, human feet are not. Of the three great apes, orangutans have long toes for more suspensory grasping capabilities (see fig. 9.21) relative to the toes of African apes. African apes, having moved to the ground as terrestrial quadrupeds, have subsequent foot modifications for weight bearing (e.g., heel-strike

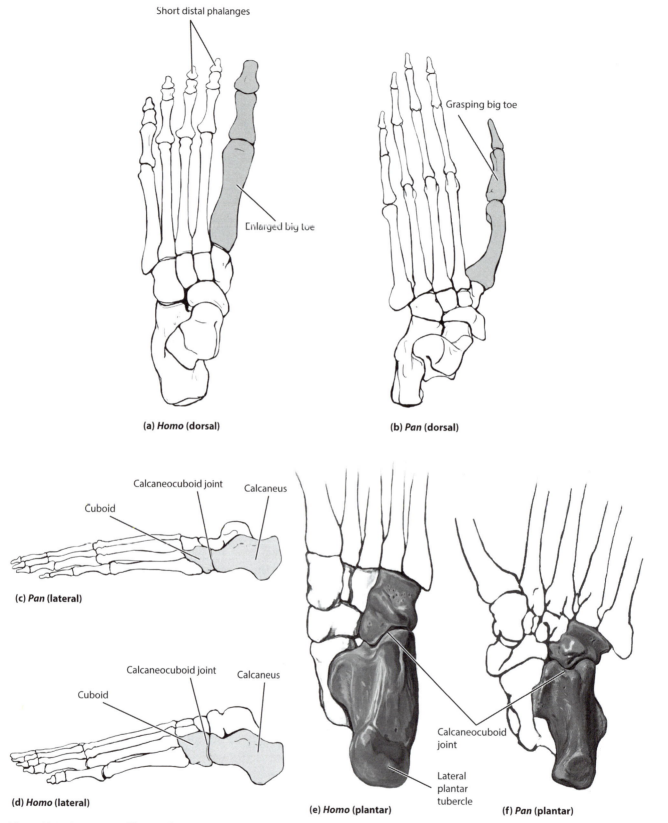

Short distal phalanges

Enlarged big toe

(a) *Homo* **(dorsal)**

Grasping big toe

(b) *Pan* **(dorsal)**

Cuboid

Calcaneocuboid joint

Calcaneus

(c) *Pan* **(lateral)**

Cuboid

Calcaneocuboid joint

Calcaneus

(d) *Homo* **(lateral)**

Calcaneocuboid joint

Lateral plantar tubercle

(e) *Homo* **(plantar)**

(f) *Pan* **(plantar)**

Figure 10.14 Great ape and human foot anatomy.

plantigrady and reduced joint mobility). African apes share more features with terrestrial humans relative to the Asian apes in this regard. Overall, our feet bear weight and are capable of little other mobility in comparison to the mobile feet and digits of great apes.

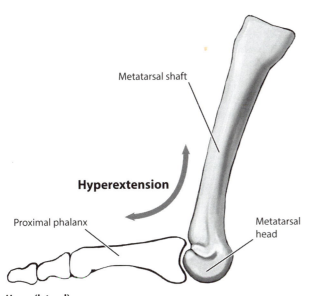

Hyperextension

Metatarsal shaft

Metatarsal head

Proximal phalanx

Homo (lateral)

Figure 10.15 Metatarsal extension in humans.

BIPEDALISM

Human adaptations for bipedalism include not only the leg and hip modifications noted above, relative to apes, but also include the loss of body heat by sweating. Humans do not pant like other mammals and we have decoupled our respiratory cycle from our locomotion. We vary our stride per breath ratios from 4:1 to 1:1 and tend to favor two strides per breath. By altering our breathing patterns and our ability to lose heat more effectively, relative to other mammals, humans possess stamina.

Dennis Bramble (University of Utah) and Daniel Lieberman (Harvard University) have suggested that humans are in fact adapted to be endurance runners. They argue that we use a compliant gait that allows tendons and muscles to store and then release energy during running. For example, human legs have long, spring-like tendons, in contrast to those of apes, and the Achilles tendon is the most important spring in the human leg. Likewise, our longitudinal arch returns energy after stance phase. The long legs of humans increase stride distance and this makes for more effective springs. Our distal leg segments are also lighter relative

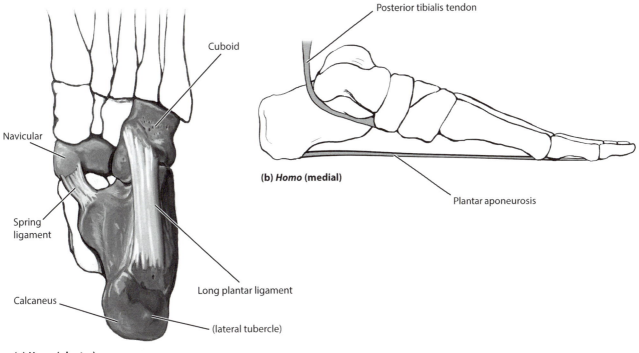

Cuboid

Navicular

Spring ligament

Calcaneus

Long plantar ligament

(lateral tubercle)

(a) *Homo* **(plantar)**

Posterior tibialis tendon

(b) *Homo* **(medial)**

Plantar aponeurosis

Figure 10.16 Elements involved in the human longitudinal arch.

With the advent of computer technology, advances in comparative anatomy followed. Morphometrics and advanced statistical approaches by Charles Oxnard led the initial wave to utilize new and sophisticated mathematical techniques to understand shape changes from a biomathematical perspective across primate anatomy. This advance was followed by in vivo bone strain gauges and electromyography (EMG) to measure forces on bone and to understand muscle activity in living primates. Both sought to supplement biomechanical approaches and concerns from an engineering approach to primate adaptation. Force plate and kinematic techniques followed to measure ground reaction forces and to study body movements across a wide range of living primates. Computer modeling (e.g., finite element analysis) can also be added to this list of advanced biomechanical approaches to better understand comparative anatomy.

Comparative anatomists also employed new imaging techniques. Scanning electron microscope images are utilized to study crown wear and enamel and dentine formation to understand primate growth and development. CT scanning (computed tomography) allows bones and fossils to be reconstructed three-dimensionally, measured, and analyzed using multivariate statistical approaches. Three-dimensional modeling techniques laser-scan bones, apply a fishnet point grid over the scan, and fit this grid to the bone. This image is then analyzed using a morphometric analysis, for example, on primate joint anatomy. All of these advanced and often quite sophisticated approaches have allowed primate comparative anatomists to examine primate bodies in new ways in our efforts to understand the phylogenetic, functional, and biomechanical significance of primate comparative anatomy (see Organ et al., 2010).

to those of African apes in that the human foot is only 9% of leg mass versus 14% in chimpanzees, suggesting a mechanically easier swing phase for humans.

As we run, peak ground forces are twice as high compared to when we walk and may approach three to four times. These high forces must be absorbed through our bones, muscles, and ligaments. Limb compliance reduces these stresses, but these forces must also be dissipated through our strong bones and larger joint surfaces (e.g., femoral head, knee joint, sacroliliac joints, lumbar centra, and the iliac pillar).

Human bodies differ from those of African apes in that we have necks with nuchal ligaments to steady our heads while running and we have wide shoulders with narrow waists to counterbalance our swinging arms and torques generated by human running and walking. In contrast, African apes do not require all of these anatomical body or limb modifications since they utilize all four limbs to support their bodies. In a similar way, the arboreal movements performed by African apes necessitate a different pattern of body and limb adaptations given the size and angled substrates African apes are capable of moving on.

Oddities

The locomotor adaptations for human bipedalism have completely remodeled the lower limbs and backs of the entire human evolutionary lineage. This is no small evolutionary feat. In fact, one might argue that

the hindlimb adaptations for human bipedalism exceed those of any other locomotor mode across primates.

CONCLUSION

Comparative anatomy is an old discipline. One of its great masters, George Cuvier (1769–1832), often considered the father of comparative anatomy, inferred that species could go extinct from his personal knowledge of comparative anatomy, a tremendous insight for science at the time. Cuvier would be shocked today given the scale of fossil discoveries, the sheer number of extinctions, and of course the impressive record of human ancestry. The field of comparative anatomy, like the work of Cuvier, has contributed fundamental insights in our understanding of biological diversity across our planet time and time again. Comparative anatomy was one of the first disciplines to actually provide empirical evidence for Linnaeus's view on biological taxonomy and later for Darwin's views concerning biological evolution. Comparative anatomy still thrives today among taxonomic and evolutionary studies. It is certainly more difficult to obtain fresh cadavers than the minute tissue samples required in other biological fields, yet the great comparative collections housed in the major natural history museums around the world offer a tremendous resource for future scientific inquiries. Comparative anatomy may even be out of fashion given the technological innovations and tremendous grant dollars awarded to other scientific fields, such as molecular biology, but given the great lists of osteological characters utilized in modern evolutionary studies, our knowledge of comparative anatomy is even more essential today than in the past.

Living primates represent a diverse group of mammals. They are dramatically different in size, in their movements and social patterns, and of course in their wide-ranging dietary preferences. As a group they represent the ancestral foundation for all human biology. Although primate studies continue and the field of primatology grows ever larger, especially in terms of scientific output, our worst fear is that our study subjects are disappearing at an ever more rapid rate of decline. The devastation of primate habitats worldwide is colossal and the levels of destruction show no decline with time or understanding. There are endangered primates in every part of the tropics. Our conservation efforts are currently formulated to limit losses since we cannot save all of the species. We are in triage mode and we may see primate icons like gorillas and orangutans go extinct within our children's lifetimes. Globally, this is a terrible environmental situation to contemplate and a scientific nightmare. We need to find solutions in which humans and primates can ecologically co-exist or face the dire consequences of a depleted planet. Extinction is forever.

Selected References

Aiello, L., and C. Dean. 1990. An Introduction to Human Evolutionary Anatomy. Academic Press, New York.

Ankel-Simons, F. 2000. Primate Anatomy—An Introduction. Academic Press, New York.

Ashton, E.H., and C.E. Oxnard. 1964. Functional adaptations of the primate shoulder girdle. Proceedings of the Zoological Society of London 142:49–66.

Bramble, D.M., and D.E. Lieberman. 2004. Endurance running and the evolution of Homo. Nature 432:345–352.

Carrier, D.R. 1984. The energetic paradox of human running and hominid evolution. Current Anthropology 25:483–495.

Gebo, D.L. 1992. Plantigrady and foot adaptation in African apes: implications for hominid origins. American Journal of Physical Anthropology 89:29–58.

———. 1996. Climbing, brachiation, and terrestrial quadrupedalism: historical precursors of hominid bipedalism. American Journal of Physical Anthropology 101:55–92.

Kapanji, I.A. 1987. The Physiology of the Joints, Vol. 2: Lower Limb. 5th edition. Churchill Livingstone, Edinburgh.

Keith, A. 1923. Man's posture: its evolution and disorders. British Medical Journal 1:451–454, 545–548, 587–590, 624–626, 669–672.

Larson, S.G. 1988. Subscapularis function in gibbons and chimpanzees: implications for interpretation of humeral head torsion in hominoids. American Journal of Physical Anthropology 76:449–462.

Latimer, B., and C.O. Lovejoy. 1989. The calcaneus of Australopithecus afarensis and its implications for the evolution of bipedality. American Journal of Physical Anthropology 78:369–386.

Latimer, B., and C.O. Lovejoy. 1990. Metatarsophalangeal joints of Australopithecus afarensis. American Journal of Physical Anthropology 83:13–23.

Latimer, B., J.C. Ohman, and C.O. Lovejoy. 1987. Talocrural joint in African hominids: implications for Australopithecus afarensis. American Journal of Physical Anthropology 74:155–175.

Le Gros Clark, W.E. 1959. The Antecedents of Man: An Introduction to the Evolution of Primates. Edinburgh Press, Edinburgh.

Lewis, O.J. 1969. The hominoid wrist joint. American Journal of Physical Anthropology 30:251–268.

———. 1989. Functional Morphology of the Evolving Hand and Foot. Oxford Science Publications, Oxford, UK.

Lovejoy, C.O. 1988. Evolution of human walking. Scientific American 259:118–125.

———. 2005a. The natural history of human gait and posture, Part 1. spine and pelvis. Gait and Posture 21:95–112.

———. 2005b. The natural history of human gait and posture, Part 2. hip and thigh. Gait and Posture 21:113–124.

———. 2006. The natural history of human gait and posture, Part 3. the knee. Gait and Posture 25:325–341.

Napier, J.R. 1967. The antiquity of human walking. Scientific American 216:56–66.

———. 1980. Hands. Pantheon Books, New York.

Organ, J.M., V.B. Deleon, Q. Wang, and T.D. Smith. 2010. From head to tail: new models and approaches in primate functional anatomy and biomechanics. The Anatomical Record 293:544–548.

Schultz, A.H. 1969. The Life of Primates. Weidenfeld and Nicolson, London.

Shea, B.T. 1985. On aspects of skull form in African apes and orangutans, with implications for hominoid evolution. American Journal of Physical Anthropology 68:329–342.

Susman, R.L. 1994. Fossil evidence for early hominid tool use. Science 265:1570–1573.

Swindler, D.R. 2002. Primate Dentition—An Introduction to the Teeth of Non-Human Primates. Cambridge University Press, Cambridge, UK.

Tuttle, R.H. 1967 Knuckle-walking and the evolution of hominoid hands. American Journal of Physical Anthropology 26:171–206.

Unger, P.S. 1995. Fruit preference of four sympatric primate species at Ketambe. International Journal of Primatology 16:221–224.

Ward, S.C., and W.H. Kimbel. 1983. Subnasal alveolar morphology and the systematic position of *Sivapithecus*. American Journal of Physical Anthropology 61:157–171.

Washburn, S.L. 1968. The Study of Human Evolution (Congdon Lectures). University of Oregon Books, Eugene.

INDEX

DANIEL L. GEBO received his Ph.D. from Duke University in 1986. After a post-doctoral fellowship at Johns Hopkins University Medical School he joined the faculty of Northern Illinois University, where he is now a professor in the Departments of Anthropology and Biological Sciences. He is a research associate at the Field Museum of Natural History (Chicago) and at the Carnegie Museum of Natural History (Pittsburgh). Professor Gebo was awarded a Presidential Research Professorship in 1998 and a Presidential Teaching Professorship in 2008 and was an inaugural winner of a Board of Trustees Professorship in 2008. He currently holds a Board of Trustees Professorship for the second time.